Prentice Hall Series in Inno...

Dennis R. Allison, David J. Farber, and Bruce D. Shriver   *Series Advisors*

A Guide to VHDL Syntax

- Based on the new IEEE Std 1076-1993

J. BHASKER

AT&T Bell Laboratories, Allentown, PA

Editorial/Production Supervision: Barbara Simpson, Amy Jolin
Acquisitions Editor: Karen Gettman
Manufacturing Manager: Alexis R. Heydt
Design Director: Jerry Votta

Prentice-Hall, Inc.
Paramount Communications Company
Englewood Cliffs, NJ 07632

The publisher offers discounts on this book when ordered in bulk quantities.
For more information, contact:

Corporate Sales Department
PTR Prentice Hall
113 Sylvan Avenue
Englewood Cliffs, NJ 07632
Phone: 201-592-2863
FAX: 201-592-2249

Printed in the United States of America

10 9 8 7 6 5 4 3 2 1

ISBN 0-13-324351-6

Prentice-Hall International (UK) Limited, London
Prentice-Hall of Australia Pty. Limited, Sydney
Prentice-Hall of Canada, Inc., Toronto
Prentice-Hall Hispanoamericana S.A., Mexico
Prentice-Hall of India Private Limited, New Delhi
Prentice-Hall of Japan, Inc., Tokyo
Simon & Schuster Asia Pte. Ltd., Singapore
Editora Prentice-Hall do Brasil, Ltda., Rio de Janeiro

*To my wife, **Geetha***

*and my two Rajahs, **Arvind** and **Vinay***

Contents

❏

Preface

VHDL is clearly becoming the de facto standard as an electronic hardware description language. It is increasingly being used to describe digital hardware designs that are applicable for both synthesis and simulation. Additionally, VHDL is commonly being used as an exchange medium, for example, between one vendor tool and another vendor tool, between an ASIC designer and the circuit user. An IEEE standard, an ANSI standard, as well as a Department of Defense standard, VHDL was first standardized by the IEEE in 1987, labeled IEEE Std 1076-1987. The language has since been modified to include new features and to reflect resolved ambiguities. This book describes the latest version of the language, labeled the IEEE Std 1076-1993.

A Guide to VHDL Syntax is the result of feedback from students in a VHDL class that I have taught for a number of years. Much of the feedback has to do with determining the syntax of a particular construct. Looking at examples from a book or class notes provides only limited usage of a construct. Reading the syntax from the Language Reference Manual (often called the LRM), the bible of VHDL, is also no easy task. It is rather cumbersome to collect all the syntax of a major construct and usually involves holding all five fingers (plus more) at various sections of the LRM to get a complete picture of the construct. This book alleviates the problem by providing the syntax of each major construct in an easy to read manner in each section. Also, the syntax of all its sub-constructs, down to such basics as identifiers and expressions, are presented within the same section in the book. Another major problem with the LRM is the lack of examples for a particular construct. In addition to describing the complete syntax of the IEEE Std 1076-1993 version of the language, this book also provides a number of examples for each construct clearly explaining how they relate to the syntax of the construct.

What is clearly lacking here are the semantics for each construct, and for this I recommend reading any good text book on VHDL, for example, the book *A VHDL Primer*, published by Prentice Hall. Further detailed semantics can be obtained by reading the LRM. I suggest the following as a complete VHDL reference set: this VHDL syntax guide, plus a VHDL book to explain semantics, plus the LRM.

This book is intended as a reference for any reader who wants to write VHDL models and wishes to see the complete syntax of a single construct in one place, with many examples.

Organization of book

The first chapter gives an introduction to the VHDL language from the syntax point of view.

Chapter 2 contains a description of all the constructs of the language, one major construct per section. Examples of many different forms of its use are also given in each section.

All major constructs are listed in the table of contents. The index contains a cross-reference for all major, as well as, minor constructs. These two mechanisms provide for fast and direct access to a specific construct.

The predefined environment in VHDL is described in Chapter 3. Many examples are given for the predefined attributes. Contents of packages STANDARD and TEXTIO are described.

Chapter 4 describes the new features in the VHDL '93 version of the language as compared to VHDL '87. This chapter also describes features in VHDL '87 that are not portable to VHDL '93.

In all the VHDL descriptions that appear in this book, reserved words are in **boldface**. Occasionally, ellipsis points (...) are used in VHDL source to indicate code that is not relevant to that discussion.

Acknowledgments

I am very grateful to the many people who have helped make this book possible. I thank Sindhu Xirasagar, Bill Paulsen, Gabe Moretti, Doug Smith, Dinesh Bettadapur, Rolf Ernst, and Micheal J. Williams, all of whom provided a much needed critical and helpful review.

I am especially indebted to Peter Ashenden for providing painstakingly detailed comments and suggestions that have resulted in a much improved book.

This book would not have been possible without the continued support and encouragement of my wife and two sons who let me steal many a precious play time from them.

J. Bhasker

April, 1994

CHAPTER 1   *Introduction*

This chapter provides a very brief introduction to the VHDL language from the syntax point of view. It also describes the necessary aspects of the host environment needed to compile and simulate VHDL models.

1.1   *About VHDL*

VHDL is a hardware description language intended for documenting and modeling digital systems ranging from a small chip to a large system. It can be used to model a digital system at any level of abstraction ranging from the architectural level down to the gate level.

The language was initially developed expressly for Department of Defense VHSIC (Very High Speed Integrated Circuits) contractors. However, due to an overwhelming need in the industry for a standard hardware description language, the VHSIC Hardware Description Language (VHDL) was selected and later approved to become an IEEE standard called the IEEE Std 1076-1987. The language was updated again in 1993 to include a number of clarifications in addition to a number of new features like groups and shared variables. This syntax guide is based on the latest version of the language called the IEEE Std 1076-1993.

1

1.2   *Modeling*

VHDL provides five kinds of design units to model a design (or a circuit):

1. Entity declaration
2. Architecture body
3. Configuration declaration
4. Package declaration
5. Package body

An *entity declaration* describes the interface of the design to its external environment, that is, it describes the ports (inputs, outputs, etc.) through which the design communicates with other designs.

An *architecture body* describes the composition or functionality of a design. This could be described as a mix of sequential behavior, concurrent behavior, and components. A design may have more than one architecture body, each describing a different composition, that is, using a different style of design. Figure 1.1 shows such a scenario.

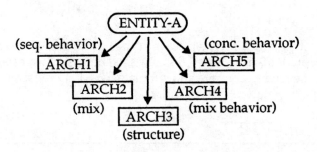

Figure 1.1 An entity with many architecture bodies.

An entity declaration and one architecture body defines a *design entity*. See Figure 1.2. The design entity represents one complete view of a design. Thus a design may have many different design entities, each representing a different view of the same design. For example, to represent a 1-bit full-adder, one entity declaration is required. In addition, there may be three architecture bodies, one that describes the behavior of the full-adder, another architecture body that describes the full-adder as a netlist of simple gates from a library such as AND and OR gates, and a third architecture body describing the full-adder using half-adders. In this case, there are three design entities for the 1-bit full-adder each one consisting of the same entity declaration. Thus, one or more design entities may share the same entity declaration but use different architecture bodies.

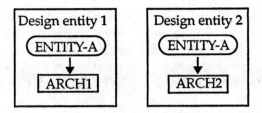

Figure 1.2 Two design entities.

A design may also be modeled at the interface level in many different ways, each of these in a different entity declaration. In the design shown in Figure 1.3, the design has three different abstractions at the external level, that is, it has three alternative interfaces. This implies one entity declaration for each interface, namely E1, E2, and E3. For example, entity E1 may have interface ports which are of type INTEGER, entity E2 may have interface ports of type BIT_VECTOR, and so on. An entity declaration with one of its corresponding architecture body represents a unique and distinct view of the design, that is, it is a design entity. In this example, the design being modeled has seven distinct design entities.

A *configuration declaration* is used to specify the bindings of components present in an architecture body to other design entities. An entire hierarchy of a design, that is, the bindings that link all the design entities in a hierarchy, can also be specified using the configuration declaration.

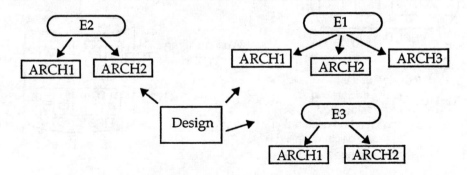

Figure 1.3 Different hardware abstractions of the same design.

A *package* is a repository to store commonly used declarations and subprograms. It is described using a *package declaration* and a *package body*. The declarations within a package can be imported into another design unit using library and use clauses.

1.3   Host environment

A VHDL *host environment* provides the capability to compile and simulate VHDL models. The environment typically supports a VHDL analyzer, a simulator, a source-level debugger, a results display viewer, and other support utilities. A *design file* is the basic unit that can be compiled by a VHDL analyzer. The design file represents a legal VHDL description and contains one or more design units.

Figure 1.4 shows a compilation scenario. Compiled descriptions, one for each design unit, are stored in design libraries. A *design library* is an implementation-specific location provided by the host environment to store compiled descriptions. VHDL allows for any number of design libraries to exist. Each design library must have a logical name. The mapping of the logical name to the actual location in the host environment is provided by the host environment. The mapping mechanism is not dictated by the language. Of all the specified design libraries, one must be designated as the working library before compilation can proceed. This capability is again provided by the host environment. A VHDL analyzer compiles each design unit from a design file and stores the compiled description for each design unit into the working library. The design library that is designated as the working library has the logical name *WORK*.

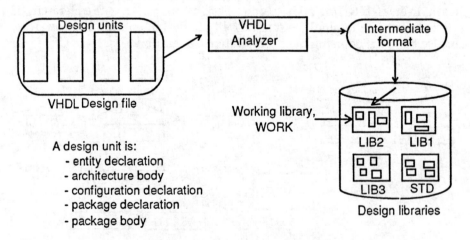

Figure 1.4 Compiling VHDL models.

The host environment also provides a predefined design library called *STD* that contains two predefined packages: *STANDARD* and *TEXTIO*. There also exists an IEEE standard package called *STD_LOGIC_1164*. This package defines a nine-value logic type, called *STD_ULOGIC*, and contains its associated subtypes, overloaded operator functions, and other useful utilities. This standard is called the IEEE Std 1164-1993. If available, a host environment must provide this package in a design library called IEEE.

Once all design units related to a complete design have been compiled, the top-level design entity or its configuration is selected for simulation. A simulator then proceeds through the following three stages:

1. Elaboration: Expands the hierarchy down to a set of behavioral models.

2. Initialization: Signals are initialized and processes are executed until they all wait.

3. Event-based simulation: Simulation starts from time 0 and simulation progresses based on events, until no more events are present, or TIME'HIGH is reached (TIME'HIGH is the maximum value for type TIME as defined in the predefined package, STANDARD).

1.4   Syntax overview

This section gives a high-level overview of the VHDL syntax. Simple terms are used to explain the format of some of the major constructs that are described in this section. Square brackets denote optional items.

A VHDL source file, or a design file, that can be compiled successfully by a VHDL analyzer contains one or more design units. A design unit is an entity declaration, an architecture body, a configuration declaration, a package declaration, or a package body.

Entity declaration

```
[ library-and-use-clauses ]
entity entity-name is
    [ generic ( list-of-generics-and-their-types ); ]
    [ port ( list-of-ports-their-type-and-mode ); ]
    [ declarations ]
[ begin
    entity-statements ]
end [ entity ] [ entity-name ] ;
```

Architecture body

```
[ library-and-use-clauses ]
architecture architecture-name of entity-name is
    [ declarations ]
begin
    concurrent-statements =>
        process-statement
        concurrent-assignment-statement
        concurrent-procedure-call
        concurrent-assertion-statement
        generate-statement
        block-statement
        component-instantiation-statement
end [ architecture ] [ architecture-name ] ;
```

Process statement

```
[ process-label : ] [ postponed ] process [ ( sensitivity-list ) ] [ is ]
    [ declarations ]
begin
    sequential-statements =>
        wait-statement
        assertion-statement
        report-statement
        signal-assignment-statement
        variable-assignment-statement
        procedure-call-statement
        if-statement
        case-statement
        loop-statement
        next-statement
        exit-statement
        return-statement
        null-statement
end [ postponed ] process [ process-label ] ;
```

Function body

```
[ pure-or-impure ] function function-name
        [ ( list-of-parameters-and-their-types ) ] return return-type is
    [ declarations ]
begin
    sequential-statements
end [ function ] [ function-name ] ;
```

Procedure body

```
procedure procedure-name [ ( list-of-parameters-and-their-types ) ] is
     [ declarations ]
begin
     sequential-statements
end [ procedure ] [ procedure-name ] ;
```

Block statement

```
block-label : block [ ( guard-expression ) ] [ is ]
     [ generic ( list-of-generics ) [ generic map ( map-values ) ] ]
     [ port ( list-of-ports ) [ port map ( map-values ) ] ]
     [ declarations ]
begin
     concurrent-statements
end block [ block-label ] ;
```

Configuration declaration

```
[ library-and-use-clauses ]
configuration configuration-name of entity-name is
     block-configuration     -- Contains binding info about components.
end [ configuration ] [ configuration-name ] ;
```

Package declaration

```
[ library-and-use-clauses ]
package package-name is
     [ declarations ]
end [ package ] [ package-name ] ;
```

Package body

```
[ library-and-use-clauses ]
package body package-name is
     [ declarations ]
end [ package body ] [ package-name ] ;
```

Detailed syntax with examples for the above constructs appear in the following chapter.

❑

CHAPTER 2   *Syntax Guide*

This chapter describes the complete syntax of the VHDL language. Examples for each construct are provided for better understanding. The constructs are arranged in alphabetical order to make referencing easier.

In all the VHDL descriptions that appear in this book, reserved words are in **boldface**. Occasionally, ellipsis points (. . .) are used in VHDL source to indicate code that is not relevant to that discussion.

General notes

VHDL is case-insensitive, that is, lowercase and uppercase letters are treated the same (except in character literals, string literals, and extended identifiers). Also, VHDL is free format, that is, a construct can appear on a single line or on multiple lines. An end of line has no significance except in a comment. The only restriction is that names and literals may not be broken across line boundaries. Here are some examples of identifiers.

```
Clock       CLOCK      clock
    -- All refer to the same identifier.
"Report"    "REPORT"
    -- These are two distinct string literals.
```

Loop   LOOP   loop
    -- All refer to the same reserved word.
\Guide\   \guide\   \GUIDE\   guide
    -- These are four distinct identifiers.

if RST = '0' **then** SDY <= '1'; **end if;**
-- is equivalent to:
IF
        rst = '0'
THEN
        sdy <= '1';
END IF;

A comment starts with two hyphens (--) anywhere on a line and ends at the end of that line

-- This line is a comment.
SUM <= A **xor** (B **xor** C);   -- This text after 2 hyphens is also
                                        -- a comment.

The root construct of a VHDL description is a "design file", that is, a design file represents a syntactically legal VHDL description.

Data objects

A *data object* holds a value of a certain type. There are four classes of data objects. These are

1. *Constant* class: An object of this class (often called a *constant*) holds a single value of a type, and the value is set when the constant is created. Once set, the value can never be changed.

2. *Variable* class: An object of this class (often called a *variable*) also holds a single value of a type, but the value of a variable can be changed using a variable assignment statement. Values are always assigned to variables immediately.

3. *Signal* class: An object of this class (often called a *signal*) holds not only the current value of a type but also the past value and the set of scheduled future values that are to appear on that signal. A signal is assigned a value using a signal assignment statement. Furthermore, a signal can be assigned a value only at a future point in time, that is, the current value of a signal can never be changed.

4. *File* class: An object of the file class (often called a *file*) contains a sequence of values of a specific type. Values can be read from a file or written to a file using read or write procedures, respectively.

Here are some examples of object declarations.

constant MAX_LINE: NATURAL := 132;

-- MAX_LINE is a constant of type NATURAL whose value is 132.

variable SEL_1: STD_LOGIC;

-- SEL_1 is a variable of type STD_LOGIC.

signal S1, S2, S3, S4: BIT_VECTOR(0 **to** 6);

-- S1, S2, S3, S4 are signals each of which is a 7-bit vector.

file ARRAY_FILE: TEXT **open** WRITE_MODE **is** "UART.array";

-- ARRAY_FILE is a file of the predefined file type TEXT, which is opened in
-- write mode, and it is linked to the physical file "UART.array" that is present
-- in the host environment (the file is created if not present).

Types and subtypes

A *type* is characterized by a set of values. There are many different types in VHDL. These are

1. Integer
2. Floating point
3. Enumeration
4. Physical
5. Array
6. Record
7. Access
8. File

Here are some examples.

type OP **is range** 0 **to** 31;

-- OP is an integer type that has the values 0 through 31.

type AMPLITUDE **is range** –1.0 **to** +1.0;

-- AMPLITUDE is a floating point type that has real values in the
-- specified range.

```
type BOOL is (NO, YES);
```
-- BOOL is an enumeration type that has the two values, NO and YES.

```
type DBUS is array (NATURAL range <>) of STD_ULOGIC;
```
-- DBUS is an unconstrained array type that has elements of
-- type STD_ULOGIC.

The predefined types of the language are described in Chap. 3.

A *subtype* is a type with a constraint. Here are some examples.

```
subtype NEW_OP is OP range 0 to 15;
```
-- NEW_OP is the subtype that is of type OP and its range is restricted
-- to 0 through 15.

```
subtype HALF_AMPLITUDE is AMPLITUDE range –0.5 to +0.5;
```
-- HALF_AMPLITUDE is a subtype of type AMPLITUDE with its range
-- restricted to –0.5 to +0.5.

```
subtype DBUS16 is DBUS(0 to 15);
```
-- DBUS16 is a subtype of type DBUS with its index range constrained
-- to 0 through 15.

About the constructs

This chapter is divided into a number of sections. A separate section is provided for each major construct in the language. Each major construct is described in terms of other major constructs and in terms of some lower-level constructs that are in turn described using some primitive constructs, whose meanings are assumed. Examples of these primitive constructs are

- identifier
- name (like slice name, attribute name)
- subtype_indication (specifies the type information)
- expression
- literals (like character literal, based literal, integer literal)

Many lower-level constructs are duplicated across many sections so that the syntax of a major construct is self-contained in that section. For example, "formal_part" and "binding_indication" appear in many sections. A construct that is marked with a superscript "mod", for instance,

formal_part^{mod}

reflects a construct that has been modified in this book to reflect its usage as it relates to the major construct that is being described in the enclosing section.

About syntax diagrams

In the syntax diagrams that follow, words in bold are keywords that have to be used as is. Some examples of keywords are

    loop   **access**   **type**

    Words in italics specify the semantics associated with that particular construct. For example,

type / subtype name	-- refers to the name of a type or a subtype.
boolean expression	-- refers to an expression that has a boolean -- value.
static expression	-- is an expression that is either locally or -- globally static.

    Arrows show the flow that is to be followed in generating the complete syntax for that construct.

2.1   *Access type declaration*

An access type represents pointers to objects of a certain type (like pointers in the C programming language). Among the various object classes, only variables can be of an access type. The default initial value of this type is the literal, null.

Every access type declaration implicitly declares a deallocation procedure called DEALLOCATE. Allocation is done using the **new** operator.

Syntax

access_type_declaration

$\longrightarrow$ **type** $\rightarrow$ identifier $\rightarrow$ **is** **access** $\rightarrow$ subtype_indication$\rightarrow$ **;** $\longrightarrow$

Used In

"Type declaration" on page 219

Examples

```
type INDEX is range 0 to 15;            -- INDEX is an integer type.
type INT_PTR_TYPE is access INDEX;

      -- INT_PTR_TYPE is an access type. It represents pointers to
      -- objects of type INDEX. The following procedure is implicitly declared
      -- from the above access type declaration:

      -- procedure DEALLOCATE (P: inout INT_PTR_TYPE);

variable A: INT_PTR_TYPE;
      -- Variable A contains an address that points to an integer.
      -- Initial value of A is null, which is its default value.

variable B: INDEX;
      -- B contains an integer value (contrast this with variable A).
      -- Initial value of B is 0.
```

```
type MEM_BLOCK is
    record
        START, STOP: NATURAL;
        USED_FLAG: BOOLEAN;
    end record MEM_BLOCK;            -- MEM_BLOCK is a record type.
type MEM_PTR_TYPE is access MEM_BLOCK;
```

-- MEM_PTR_TYPE is an access type that represents addresses that point to
-- objects of type MEM_BLOCK. The following deallocation procedure is
-- implicitly declared:

-- **procedure** DEALLOCATE (P: **inout** MEM_PTR_TYPE);

variable C: MEM_PTR_TYPE := **new** MEM_BLOCK'(25, 48, FALSE);

-- Variable C contains an address that points to a record of type
-- MEM_BLOCK with its initial values as specified, i.e. START has value 25,
-- STOP has value 48, and USED_FLAG is FALSE.

2.2   *Aggregate*

An aggregate is a collection of expressions that represent a composite value of an array or a record type.

   If an expression is specified with a choice (in this context, a choice is an index or a range of indices), named association is said to be used to specify the values, else if no choice is specified, positional association is said to be used.

Syntax

aggregate

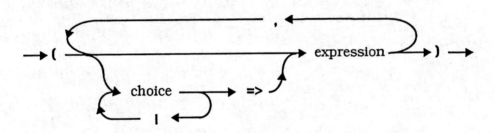

choice

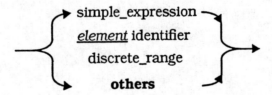

discrete_range

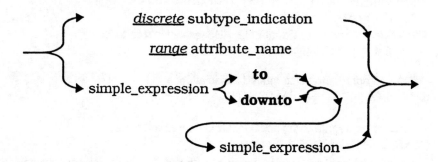

Used In

"Conditional signal assignment statement" on page 67
"Qualified expression" on page 177
"Selected signal assignment statement" on page 184
"Signal assignment statement" on page 189
"Simple expression" on page 195
"Variable assignment statement" on page 225

Examples

```
-- Array aggregates using positional association:
('0', '0', '0', '1', '1')
(SUM, CARRY)
(A0, A1, A2, A3)

-- Here is an example of a usage of an array aggregate:
variable SHREG: BIT_VECTOR(5 to 9);
SHREG := ('0', '1', '0', '1', '1');
    -- This implies that bit 5 of variable SHREG gets the value '0',
    -- SHREG(6) gets '1', and so on.

-- Array aggregates using named association:
(0 | 1 | 5 => 'U', others => '1')
(0 to 3 => '0', 4 to 11 => '1', 12 to 15 => '0')
(others => '0')
(10 | 8 => SR, 5 => DY, others => KS)
    -- The keyword others, if present, must be the last element in the aggregate.
```

-- Here is an example of a usage of an array aggregate:
signal EM2_ABS: STD_LOGIC_VECTOR(7 **downto** 0)
    := (0 | 1 | 5 => 'U', **others** => '1');
    -- Bits 0, 1, and 5 of signal EM2_ABS are initialized to 'U',
    -- and all remaining bits are initialized to '1'.

-- A record aggregate using named association:
(DAY => 26, MONTH => JANUARY, YEAR => 1993)

-- A record aggregate using positional and named association:
("DATA", "INTEGER", 26, **others** => 5)
    -- All positional associations must appear before any named associations.

-- An array aggregate using positional and named association:
('1', '0', '1', COIN'RANGE => 'U', **others** => 'Z')
    -- All positional associations must appear before any named associations.

-- Here is an example of a usage of a record aggregate:
type MONTH_TYPE **is** (JAN, FEB, MAR, APR, MAY, JUN, JUL,
                        SEP, OCT, NOV, DEC);
type DATE **is**
    record
        YEAR   : NATURAL;
        MONTH : MONTH_TYPE;
        DAY   : INTEGER **range** 0 **to** 31;
    end record;
variable READMUX: DATE;
READMUX := (DAY => 26, MONTH => JAN, YEAR => 1993);
    -- The DAY field of record READMUX gets value 26, the MONTH field
    -- gets the value JAN, and the YEAR field gets the value 1993.

2.3   Alias declaration

An object (that is, a variable, constant, signal, or a file) or a non-object (that is, any other named item such as subprogram name, enumeration literal, or type, except for labels, loop and generate parameters) can be assigned an alternate name, that is, an alias, using an alias declaration. The alias and the aliased name can be used interchangeably in a VHDL description. A parameter profile, called the *signature*, is required if the alias is for a subprogram or an enumeration literal.

A subtype indication is not allowed for a non-object alias. A signature is only allowed for a subprogram or an enumeration literal alias.

Syntax

alias_declaration

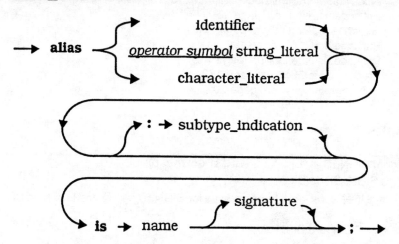

signature

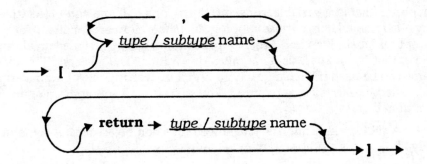

Used In

"Block statement" on page 42
"Entity declaration" on page 95
"Generate statement" on page 120
"Package body" on page 159
"Package declaration" on page 162
"Process statement" on page 172
"Subprogram body" on page 202

Examples

signal SIGNED_A: BIT_VECTOR(31 **downto** 0);
alias SIGN: BIT **is** SIGNED_A(31);
    -- SIGN is declared to be an alias for SIGNED_A(31) and is of type BIT.

BAX <= SIGN **and** B(31);
    -- The right hand side expression is equivalent to
    -- writing "SIGNED_A(31) **and** B(31)".

constant CONTROL: MVL_VECTOR(0 **to** 63);
alias OP_CODE: MVL_VECTOR(4 **downto** 0) **is** CONTROL(0 **to** 4);
    -- Note range direction is changed from **to** to **downto** as a result of the alias.

alias OPD1: MVL_VECTOR(0 **to** 7) **is** CONTROL(5 **to** 12);
alias OPD2: MVL_VECTOR(0 **to** 7) **is** CONTROL (13 **to** 20);
alias ADDR: MVL_VECTOR(0 **to** 43) **is** CONTROL (20 **to** 63);

-- In the function call,

    . . . COMPUTE (OPD1, OPD2, OP_CODE) . . .

-- OPD1 is equivalent to writing CONTROL(5 **to** 12), and so on.

alias "+" **is** SYNTH.BIT_ARITH."+" [SYNTH.BIT_ARITH.SIGNED,
SYNTH.BIT_ARITH.SIGNED **return** SYNTH.BIT_ARITH.SIGNED];

-- Note the use of a signature to identify the "+" overloaded function.

alias CMP65 **is** TI_CMOS65.COMP_DECL;

-- CMP65 is an alternate name for the package "TI_CMOS65.COMP_DECL".

alias VLO **is** ZCD.LOGIC_PK.'0' [**return** ZCD.LOGIC_PK.LOGIC7];

-- VLO is an alias for the character literal '0' present in the
-- package LOGIC_PK that resides in the design library ZCD.

-- CMP65, "+", VLO are non-object aliases.
-- SIGN, OP_CODE, OPD1, OPD2, ADDR are object aliases.

2.4   *Allocator*

An allocator, when executed, creates a variable object of the specified type and returns the pointer to the object. This is similar to the function "malloc" in C programming language. The value of the variable object is the same as the initial value of a variable of that type, or it can be explicitly specified using a qualified expression.

The value of the variable can be accessed by dereferencing the pointer. If *obj* is the pointer that points to a variable, then

- *obj*.all gives the value of the variable.
- *obj* [*index*] gives the value of the array element.
- *obj.element-name* gives the value of the record element.

Syntax

allocator

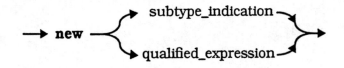

Used In

"Simple expression" on page 195

Examples

type INDEX **is** 10 **to** 15;
-- INDEX is an integer type.

type INDEX_PTR **is access** INDEX;
-- INDEX_PTR is an access type.

variable T1, T2: INDEX_PTR;
T1 := **new** INDEX;
-- T1 is a pointer that points to a variable object of type INDEX whose initial
-- value is 10. The expression "T1.**all**" will give the value 10.

T2 := **new** INDEX'(14);

    -- T2 is a pointer that points to a variable object of type INDEX and whose
    -- initial value is set to 14.

type STD_LOGIC_PTR **is access** STD_LOGIC_VECTOR;
variable T3: STD_LOGIC_PTR :=
                    new STD_LOGIC_VECTOR'("1101");

    -- T3 is a pointer to a vector of size 4 bits.

type BIT_PTR_TYPE **is access** BIT_VECTOR;
variable BIT_PTR: BIT_PTR_TYPE := **new** BIT_VECTOR'("00101");
variable BIT_PTR2: BIT_PTR_TYPE;
BIT_PTR2 := **new** BIT_VECTOR;

    -- This is illegal since the type BIT_VECTOR is not constrained.

BIT_PTR2 := **new** BIT_VECTOR (5 **downto** 0);

    -- BIT_PTR points to a 6-bit string with the default value of "000000".

type INT_LIST_NODE;

    -- This is an incomplete type declaration. It is needed so that the type
    -- can be used in the following access type declaration.

type INT_LIST_PTR **is access** INT_LIST_NODE;
type INT_LIST_NODE **is**
    record
        INT_VALUE: POSITIVE;
        NEXT_PTR: INT_LIST_PTR;
    end record;   -- This is the complete type declaration.

variable INT_LIST: INT_LIST_PTR;
INT_LIST := **new** INT_LIST_NODE;

    -- INT_LIST points to a record whose fields have the default values,
    -- 1 for INT_VALUE and **null** for NEXT_PTR.

INT_LIST := **new** INT_LIST_NODE'(INT_VALUE => 22,
                           NEXT_PTR => INT_LIST);

    -- INT_LIST points to a record whose field values are explicitly specified
    -- using named association. Note that NEXT_PTR points to the previous
    -- value of INT_LIST, not to the INT_LIST value on the left hand side.

2.5   *Architecture body*

An architecture body describes the internal view of a design entity. This can be expressed using any of the following styles:

1. *Behavioral style*: Describes the behavior of the design entity as a sequential program. This is accomplished by using sequential statements within a process.

2. *Dataflow style*: Describes the behavior as a set of concurrent statements. This is done primarily using concurrent signal assignment statements and block statements.

3. *Structural style*: Describes the structure of the design entity as a set of interconnected components. This is done by using component instantiation statements.

4. Any mix of the above.

An architecture body is always associated with an entity declaration. All declarations that appear between the keywords **is** and **begin**, called the architecture declarative part, declare local names that are visible to all statements inside the architecture body. For example, signals may be declared here which are visible to all statements inside the architecture body. Concurrent statements are a set of statements that execute concurrently with respect to each other within an architecture body.

Syntax

architecture_body

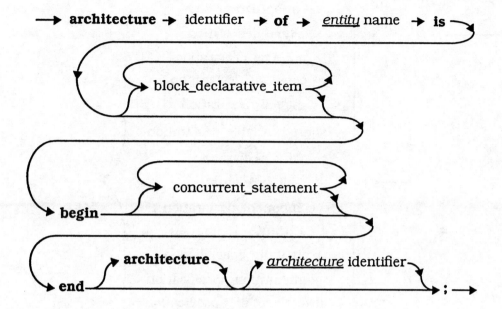

block_declarative_item

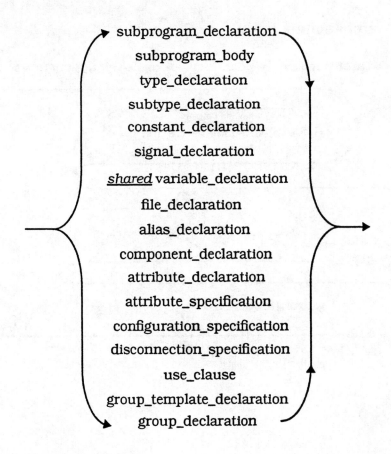

subprogram_declaration
subprogram_body
type_declaration
subtype_declaration
constant_declaration
signal_declaration
shared variable_declaration
file_declaration
alias_declaration
component_declaration
attribute_declaration
attribute_specification
configuration_specification
disconnection_specification
use_clause
group_template_declaration
group_declaration

Used In

"Design file" on page 90

Examples

```
architecture ARCHITECTURE_NAME of ENTITY_NAME is
    -- Declarations that appear here are local to the architecture body.
begin
    -- Concurrent statements here.
end architecture ARCHITECTURE_NAME;
```

```
architecture ONE_VIEW of FFT is
    -- ONE_VIEW is the name of the architecture.
    -- FFT is the name of the entity. There are two local declarations in
    -- the architecture declarative part:
    signal S1: BIT;
    constant XOR_DELAY: TIME := 5 ns;
begin
    -- Following are two concurrent statements, assigning the output SUM
    -- based on inputs, A, B, C:
    S1 <= A xor B after XOR_DELAY;
    SUM <= S1 xor C after XOR_DELAY;
end architecture ONE_VIEW;


-- An architecture MIXED of a full-adder entity FA_MIX using all three
-- styles of description:
architecture MIXED of FA_MIX is
    -- There are 2 local declarations in this architecture body:
    component MY_XOR is
        port (L, R: in BIT; M: out BIT);
    end component;      -- Component declaration.
    signal S1: BIT;         -- Signal declaration.
begin
    -- There are three concurrent statements in this architecture body:
    X1: MY_XOR port map (L => A, R => B, M => S1);
        -- Component instantiation statement.

    P1: process (A, B, CIN) is          -- Beginning of process statement.
        variable TEMP1, TEMP2, TEMP3: BIT;
    begin
        TEMP1 := A and B;
        TEMP2 := B and CIN;
        TEMP3 := A and CIN;
        COUT <= (TEMP1 or TEMP2) or TEMP3;
    end process P1;                      -- End of process statement.

    L1: SUM <= S1 xor CIN;        -- Concurrent signal assignment statement.
end architecture MIXED;
```

architecture DUMMY **of** DUMMY **is**
    -- Architecture name can be same as entity name.
begin
    A <= B /= C;
    -- Assign to signal A, the value of the B /= C comparison operation.
end;
    -- Keyword **architecture** and architecture name are optional after
    -- keyword **end.**

2.6   Assertion statement

An assertion statement, when executed, checks the value of a boolean expression. If the value is FALSE, a string is printed to the output, and a severity level is reported to the simulator for further action. If the value of the boolean expression is TRUE, no action takes place.

Syntax

assertion_statement

assertion

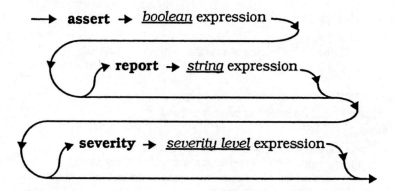

Used In

"Sequential statement" on page 188

Examples

A1: **assert** (CLK = '0') **and** DIN'STABLE (3 ns)
    report "Setup time too short!"
    severity FAILURE;

      -- This assertion statement has a label A1. If the value of the
      -- boolean expression "(CLK = '0') **and** DIN'STABLE (3 ns)" is
      -- FALSE, then the message "Setup time too short!" is printed to
      -- the standard output, and the severity level of FAILURE is passed
      -- to the simulator.

A2: **assert** DATA <= X"FF0F"
    report "DATA value exceeds FF0F(hex)!";

      -- Default severity level of ERROR is reported to simulator.

constant MISMATCH: STRING
                    := "The values of COUNT do not match.";
assert COUNT_EXPECTED = COUNT_RESULT
    report MISMATCH;

      -- This assertion statement has no label.

assert CLK='0' **or** CLK = '1';

      -- No report string is explicitly specified. If assert expression evaluates to
      -- FALSE, the default string "Assertion violation" is printed.
      -- The default severity level of ERROR is also reported to simulator.

assert FALSE
    report "This statement has been reached."
    severity NOTE;

      -- Since the assertion condition is always false, the report message
      -- is printed every time this assertion statement is executed.
      -- A report statement could also have been used instead of the above
      -- assertion statement such as:

    report "This statement has been reached.";

assert RAM_ADDR <= TO_STDLOGICVECTOR(UPPER_BOUND)
    report "RAM address out of bounds."
    severity WARNING;

2.7   *Attribute declaration*

An attribute declaration is used to declare a user-defined attribute. An attribute is a value (like a characteristic) that can be associated with certain items such as entity, architecture, process, or component label. An item is associated with an attribute with a given value using an attribute specification.

Syntax

attribute_declaration

$\longrightarrow$ **attribute** $\rightarrow$ identifier $\rightarrow$: $\rightarrow$ *type/subtype* name $\rightarrow$; $\longrightarrow$

Used In

"Architecture body" on page 24
"Block statement" on page 42
"Entity declaration" on page 95
"Generate statement" on page 120
"Package declaration" on page 162
"Process statement" on page 172
"Subprogram body" on page 202

Examples

attribute BUILT_IN: BOOLEAN;
    -- Declares an attribute called BUILT_IN that can hold only a BOOLEAN
    -- value.

attribute MAX_RISE_TIME: REAL;

attribute DONT_TOUCH: BOOLEAN;

attribute CLOCK_PERIOD: TIME;

2.8   Attribute name

An attribute name represents the value of an attribute associated with an item such as an array, entity, architecture, or label. An attribute declaration is used to declare an attribute. An attribute specification is used to specify the value of the attribute and to associate it with an item. An attribute name is used to retrieve the value of the attribute. The attribute may be a predefined attribute or a user-defined attribute.

Syntax

attribute_name

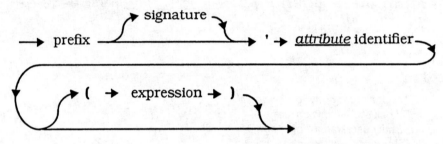

prefix

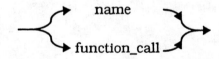

signature

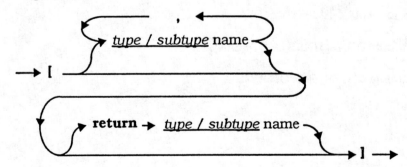

Used In

Examples

ADDR ' RIGHT

    -- ADDR is the name of an array and 'RIGHT is a predefined attribute of an
    -- array. The name, ADDR'RIGHT refers to the rightmost index of
    -- the ADDR array.

CLK_A ' DELAYED (15 ns)

    -- CLK_A is a signal name, 'DELAYED is a predefined attribute with an index
    -- value of 15 ns. This attribute name refers to a new implicit signal that has
    -- the same waveform as CLK_A but delayed by 15 ns.

COMPLEMENT_BITS (A_WORD) ' LENGTH

    -- This name refers to the value of the predefined array attribute 'LENGTH
    -- that is applied to the result of a function call that returns an array object.

"*" (A, B) [STD_LOGIC_VECTOR, STD_LOGIC_VECTOR
       return STD_LOGIC_VECTOR] ' BUILT_IN

    -- Refers to the value of the user-defined attribute 'BUILT_IN associated with
    -- the overloaded function for "*" with the specified signature.

RAM_ADDR ' REVERSE_RANGE

    -- RAM_ADDR is an array and 'REVERSE_RANGE is a predefined array
    -- attribute. This attribute name returns the array range of
    -- RAM_ADDR in reverse order.

PRESET_CLEAR [SIGNED, INTEGER] ' BUILT_IN
    -- Refers to the value of the attribute 'BUILT_IN associated with the
    -- procedure PRESET_CLEAR with the specified signature.

2.9   *Attribute specification*

An attribute specification is used to associate a user-defined attribute with a named item, such as an entity, architecture, or a label, and to specify a value for the attribute.

The attribute must have been declared earlier using an attribute declaration. The value of the attribute, which is specified using an attribute specification, may be accessed by using the attribute name.

Syntax

attribute_specification

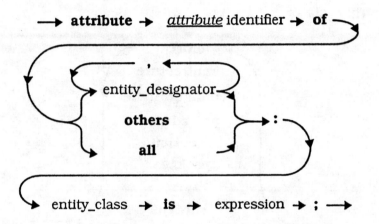

entity_designator

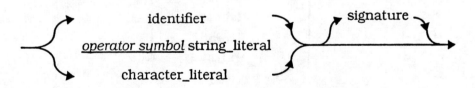

signature

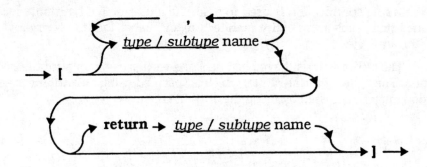

entity_class

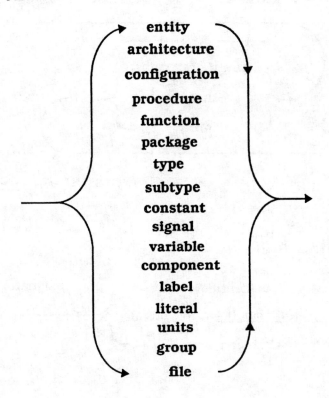

Used In

"Architecture body" on page 24
"Block statement" on page 42
"Configuration declaration" on page 71
"Entity declaration" on page 95
"Generate statement" on page 120
"Process statement" on page 172
"Subprogram declaration" on page 208

Examples

attribute BUILT_IN **of** PRESET_CLEAR: **procedure is** TRUE;

    -- BUILT_IN is the name of attribute that is being associated with the
    -- procedure PRESET_CLEAR and the value of the attribute is set to TRUE.
    -- The value of the attribute can then be accessed using an attribute name,
    -- for example:

        PRESET_CLEAR'BUILT_IN
    -- which has the value TRUE.

attribute DELAY_WIDTH **of all: variable is** 10 ns;

    -- Attribute DELAY_WIDTH is associated with all variables in that declarative
    -- region where this specification appears.

attribute MAX_RISE_TIME **of** S1, S2, S3: **signal is** 5 ns;

attribute ENUM_ENCODING **of** READY [**return** MC_STATE]: **literal
is** "001010";

    -- READY is an enumeration literal of type MC_STATE and it is associated
    -- with an attribute ENUM_ENCODING with value "001010".

attribute EXPAND **of** "+", "*": **function is** TRUE;

    -- The attribute EXPAND is being associated with all overloaded functions for
    -- "+" and "*", since no signature is specified.

attribute BUILT_IN **of** "–" [STD_LOGIC_VECTOR,
    STD_LOGIC_VECTOR **return** STD_LOGIC_VECTOR]: **function is** TRUE;

attribute DONT_TOUCH **of** C9, C4, C12, C1: **label is** TRUE;

    -- The attribute DONT_TOUCH is attached to the specified labels and has the
    -- value of TRUE.

attribute DONT_TOUCH **of others: label is** FALSE;

    -- The DONT_TOUCH attribute is attached to all other labels declared in the
    -- declarative region in which this attribute specification occurs.

attribute ENUM_TYPE_ENCODING **of** 'U' **[return** STD_ULOGIC **] :**
literal is "SYNTHESIS_UNKNOWN";

    -- The attribute ENUM_TYPE_ENCODING is attached to the enumeration
    -- literal 'U' that appears in the type declaration of STD_LOGIC and the
    -- attribute has the value of "SYNTHESIS_UNKNOWN".

2.10   Based literal

A based literal specifies an integer or a floating point number in a base system possibly other than 10. Legal value of a base ranges from 2 to 16. The integer value preceding the # symbol specifies the base, and the value between the two # symbols represents the value in that base. The exponent, if present, specifies the power to which the base is raised. The underline character ( ) has no significance and may be used for clarity.

Syntax

based_literal

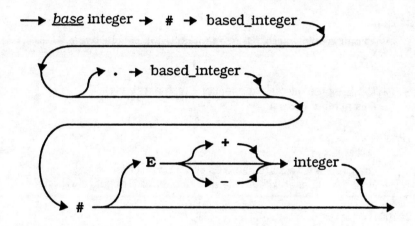

based_integer

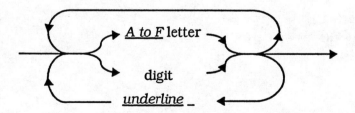

Used In

"Literal" on page 150
"Physical literal" on page 165

Examples

2#11_00#

--The integer value, 12, written in binary form.
-- Underlines are ignored and are used for clarity.

16#F_3_0_B#

-- The integer value, $(15*16^3+3*16^2+0*16^1+11*16^0=62219)$,
-- written in hexadecimal form.

8#3.24#

-- A floating point value, $(3+2/8+4/64=3.3125)$, in octal base.

4#3#E2

-- The E represents an exponent of 4 (same as base).
-- This number is $3 * 4^2$

12#a09#e5

-- The value $(A09)_{12} * 12^5$
-- Lower case letters are same as upper case letters.
-- Note that a based integer literal cannot have a negative exponent.

2#100110#

-- The integer value, 38.

16#FFF#

-- The integer value, 4095.

8#70#E+2

-- The value $(70)_8*8^2$

2.11   Bit string literal

A bit string literal represents a sequence of bit values, that is, of '0' and '1' values. It can be expressed in binary form, in octal form, or in hexadecimal form. The underline character has no significance and may be used for clarity.

Syntax

bit_string_literal

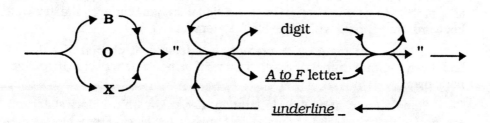

Used In

"Literal" on page 150

Examples

    B"001_101_010"   -- A binary bit string literal of length 9 bits.

    X"A_F0_FC"   -- A hexadecimal bit string literal of length 20 bits.

    O"3701"   -- An octal bit string literal of length 12 bits.

    X" "   -- A null bit string literal, that is, length is 0.

    B"0001"   -- A binary bit string literal of length 4 bits.

    X"BC0"   -- A hexadecimal bit string literal of length 12 bits.

    O"36"   -- An octal bit string literal of length 6 bits.

    O" "   -- A null bit string literal, length is 0.

2.12   Block statement

A block statement is a concurrent statement that contains other concurrent statements. It can be used for three main purposes:

1. To disable a guard
2. To partition a description
3. To limit scope of declarations

If a guard expression is present in a block statement, a signal called GUARD is implicitly declared within the block statement, and it always has the value of the guard expression. The guard expression controls any guarded assignments (a guarded assignment is a signal assignment with the **guarded** keyword) that appear within the block statement.

A block statement can represent an independent portion of a design and it can communicate with its environment using its own port map and generic map.

Signals, types, and other declarations declared within a block statement are local to that block and are not visible outside the block. Every block statement must have a label.

Syntax

block_statement

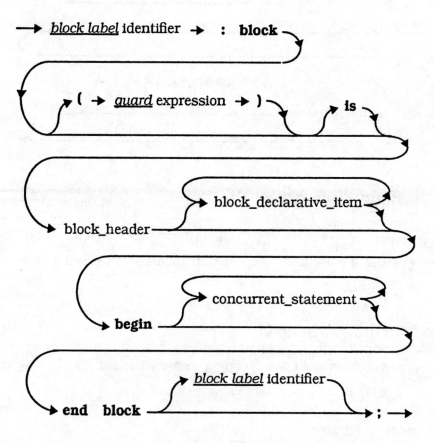

block_header

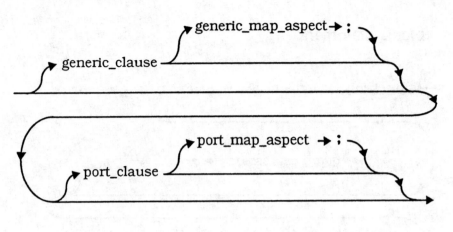

generic_clause^{mod}

→ **generic** **(** → interface_constant_declaration → **)** **;** →

generic_map_aspect

→ **generic** **map** **(** → *generic* association_list → **)** →

port_clause^{mod}

→ **port** **(** → interface_signal_declaration → **)** **;** →

port_map_aspect

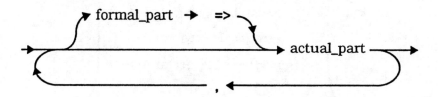

$\longrightarrow$ **port map (** $\rightarrow$ _port_ association_list $\rightarrow$ **)** $\longrightarrow$

association_list

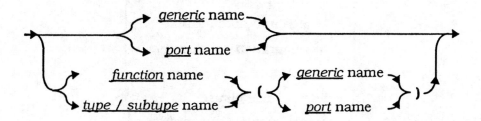

formal_part^{mod}

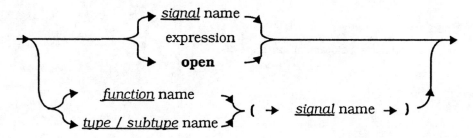

actual_part^{mod}

block_declarative_item

subprogram_declaration
subprogram_body
type_declaration
subtype_declaration
constant_declaration
signal_declaration
shared variable_declaration
file_declaration
alias_declaration
component_declaration
attribute_declaration
attribute_specification
configuration_specification
disconnection_specification
use_clause
group_template_declaration
group_declaration

Used In

"Concurrent statement" on page 66

Examples

B_ONE: **block** (CLOCK = '1') **is**
    -- Every block must have a label. This block has the label B_ONE.
    -- Since a guard expression is present, the implicit signal GUARD
    -- of type BOOLEAN always has the value of this guard expression.
    signal TEMP: BIT;
        -- Signal TEMP is local only to this block.
begin
    TEMP <= **guarded** D;   -- Guarded assignment.
    Q <= TEMP:
    QBAR <= **not** TEMP;
end block B_ONE;
    -- When an event occurs on signal D or CLOCK, the assignment to TEMP is
    -- made only if the value of the guard expression is TRUE, that is, CLOCK is
    -- '1'. If the guard expression is false, TEMP retains its old value. Note that
    -- the guard expression has no effect on the other two concurrent signal
    -- assignment statements, since they are not guarded assignments.

TB: **block**   -- The keyword **is** is optional.
    -- This block has no guard expression.
    generic (RISE, FALL: TIME);
    generic map (FALL => 4 ns, RISE => 5 ns);
        -- Named association used in generic map.
    port (A, B: **in** BIT; Z: **out** BIT);
    port map (S1, RDY, CTRL);
        -- Positional association used in port map.
    signal LOCAL: BIT;
begin
    LOCAL <= A **nand** B;
    Z <= LOCAL **after** RISE **when** (A'EVENT **and** A = '1') **or**
                         (B'EVENT **and** B = '1') **else**
        LOCAL **after** FALL;
end block;
    -- Block label is optional after **end block.**

SELECTOR: **block**
    port (M1, M2, SEL: **in** STD_LOGIC; MZ: **out** STD_LOGIC);
    port map (M1 => '0', M2 => CASE_VAL, SEL => CTRL,
        MZ => SUM);
        -- Actuals in a port map may be expressions (globally static), but their
        -- corresponding formals must be of mode **in.**

```
begin
    MZ <=   M1 when SEL = '0',
            M2 when SEL = '1',
            'U' when others;
end block SELECTOR;

SR: block
    constant NAND_DELAY: TIME := 7 ns;
begin
    Q <= S nand QBAR after NAND_DELAY;
    QBAR <= R nand Q after NAND_DELAY;
end block;
        -- Signals R, Q, S, and QBAR are declared outside of the block.

NON_INV_BUFF: block
    component INV is
        generic (TPHL, TPLH: TIME);
        port (A: in STD_LOGIC; B: out STD_LOGIC);
    end component INV;
        -- A block can contain component declarations as well.
    signal TEMP: STD_LOGIC;
begin
    V1: INV generic map (2 ns, 5 ns) port map (CLKSTATIN, TEMP);
    V2: INV generic map (7 ns, 10 ns) port map (TEMP, CLKSTATOUT);
    -- Signals CLKSTATIN and CLKSTATOUT are declared outside the block.
end block;
        -- Block statement used to decompose and represent a part of a
        -- design in structural form.

DECODER: block
    constant RAM_WRITE_DELAY: TIME := 5 ns;
    signal GUARD: BOOLEAN;
begin
    GUARD <= CLOCK = '1' and (not CLOCK'STABLE);
    RAM (RAM_ADDR) <= guarded RAM_DATA
                        after RAM_WRITE_DELAY;
end block;
        --A block statement need not have a guard expression. A signal GUARD
        -- could be explicitly declared and assigned a value within the block
        -- statement. Any guarded assignment within the block statement will then be
        -- controlled by this GUARD signal. In this example, the assignment to RAM
        -- is controlled by the guard expression, that is, assignment will occur only at
        -- the rising edge of the clock.
```

2.13   Case statement

The case statement is a sequential statement that, when executed, evaluates the select expression, compares this value with the choices, and executes the set of statements in the branch corresponding to the matching choice. Note that there is no fall-through case, as in the C programming language, in VHDL.

All values of the select expression must be covered in a case statement. The keyword **others** may be used as a catch-all for choice values that are not specified in the choices of preceding branches. Note also that each value can be specified exactly once as a choice in a case statement.

Syntax

case_statement

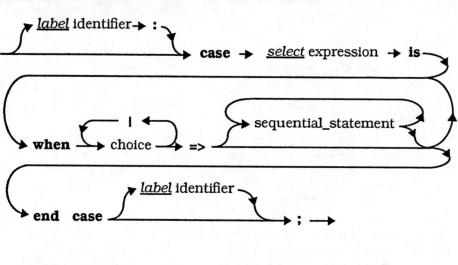

choicemod

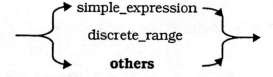

discrete_range

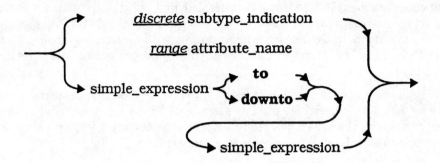

Used In

"Sequential statement" on page 188

Examples

```
signal SLEEP: MY_TIME;
...
type DAY is (SUN, MON, TUE, WED, THU, FRI, SAT);
variable TODAY: DAY;
...
C1: case TODAY is
    -- C1 is the case statement label. TODAY is the select expression.
    when SAT | SUN => SLEEP <= 10 hours;
        -- When SAT or SUN.
    when MON => SLEEP <= 6 hours;
    when TUE to THU => SLEEP <= 8 hours;
        -- When TUE, WED or THU.
    when others => SLEEP <= TGIF;
        -- When FRI.
end case C1;

type MC_TYPE is (WAITS, READY, APPLY);
signal MC_STATE: MC_TYPE;
...
case MC_STATE is                          -- This case statement has no label.
    when WAITS => if RD = '0' then
                        MC_STATE <= READY;
```

```
                              end if;
         when READY => if WR = '1' then
                              MC_STATE <= APPLY;
                         else
                              MC_STATE <= WAITS;
                         end if;
         when APPLY => if RD = '1' then
                              MC_STATE <= WAITS;
                         end if;
         -- No others clause needed since all values of the select expression
         -- are covered.
end case;

variable TOGGLE: BIT;
. . .
T1: case TOGGLE is
     when '0' => PING;
     when '1' => PONG;
end case;
         -- PING and PONG are procedures declared elsewhere.
         -- T1 is the case statement label.

type MVL is ('X', '0', '1', 'Z');
type MVL_VECTOR is array (NATURAL range <>) of MVL;
signal ODD_CTR: MVL_VECTOR(0 to 1);
. . .
case ODD_CTR is
     when "01" | "00" | "10" => ODD_CTR <= ODD_CTR + 1;
     when "11" => ODD_CTR <= "00";
     when others => report "ODD_CTR has non-binary values";
end case;
         -- In this example, it appears that all possible values have been covered by
         -- the first two when clauses. But that's not completely true since ODD_CTR
         -- can have element values other than '0' and '1', namely, 'X' and 'Z'. It is
         -- assumed that "+" has been overloaded to operate on MVL_VECTOR and
         -- integer operands.
```

2.14   *Character literal*

A character literal is a legal graphic character written within single quotes.

Syntax

character_literal

$$\longrightarrow \text{'} \to \text{graphic\_character} \to \text{'} \longrightarrow$$

Used In

"Alias declaration" on page 19
"Attribute specification" on page 35
"Enumeration literal" on page 100
"Group declaration" on page 125
"Selected name" on page 182

Examples

'A'	-- Character 'A'
'a'	-- Not same as 'A'
'&'	
'('	
'<'	
'0'	-- This is not integer 0.
'9'	
' '	-- Space character.
''''	-- The ' character itself.
'"'	-- The " character.

2.15   *Component declaration*

A component declaration declares the interface of a component. A component when instantiated in an architecture body must be first declared using a component declaration.

Syntax

component_declaration

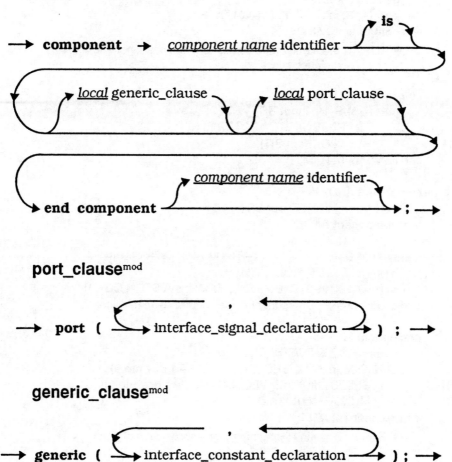

port_clause^{mod}

generic_clause^{mod}

Used In

"Architecture body" on page 24
"Block statement" on page 42
"Generate statement" on page 120
"Package declaration" on page 162

Examples

```
component XOR2 is
    generic (RISE, FALL: TIME);
    port (A, B: in X01Z; Z: out X01Z);
end component XOR2;
```
-- XOR2 is the name of the component that has a generic clause
-- and a port clause specified.

```
component RAM is
    port (DBUS_IN: inout BIT_VECTOR(0 to 7);
        ADDRESS: in BIT_VECTOR (0 to 15);
        RW, CLOCK: in BIT);
end component RAM;
```

```
component AND2 is
    port (A, B: in BIT; Z: out BIT);
end component AND2;
```

```
component DFF                  -- The keyword is, is optional.
    generic (SETUP, HOLD: TIME);
    port (D, CK: in STD_LOGIC; Q, QBAR: out STD_LOGIC);
end component; -- Component name is optional on this line.
```

```
component MUX is
    generic (SIZE: POSITIVE);
    port (DIN: in X01Z_VECTOR(2**SIZE – 1 downto 0);
        SELECT: in X01Z_VECTOR (SIZE – 1 downto 0);
        MUXOUT: out X01Z);
end component MUX;
```
-- Expressions may appear in range specification of ports.

```
component DUMMY
end component;
```
-- A component with no generic clause and no port clause.

```
component ADDER is
    port (A, B: in BIT_VECTOR; C: out BIT_VECTOR);
end component ADDER;
    -- Port types can be unconstrained arrays.
```

2.16   Component instantiation statement

A component instantiation statement represents an instance of a component in the architecture body in which it appears. A component must first be declared using a component declaration before it can be instantiated.

A component instance can be bound to a design entity that resides in a design library using a configuration specification or a configuration declaration. A component instantiation statement may also directly instantiate a design entity or a configuration.

Syntax

component_instantiation_statement

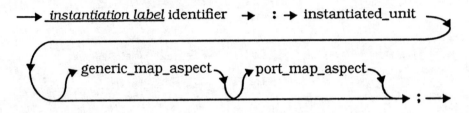

instantiated_unit

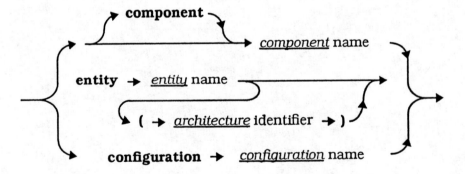

generic_map_aspect

⟶ **generic map (⟶** _generic_ association_list **⟶)** ⟶

port_map_aspect

⟶ **port map (⟶** _port_ association_list **⟶)** ⟶

association_list

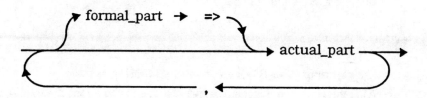

formal_partᵐᵒᵈ

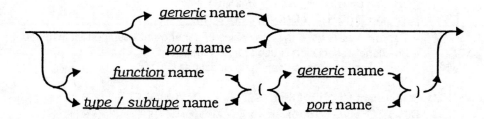

actual_part[mod]

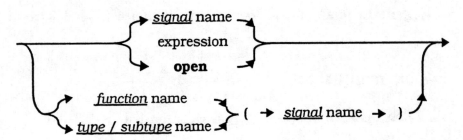

Used In

"Concurrent statement" on page 66

Examples

A1: AND2 **port map** (A => S0, B => **open**, C => COUT);

    -- A1 is the component instance label, AND2 is the component name. A, B, C
    -- are the port names of the component. Signal S0 is connected to port A, no
    -- signal is connected to port B and signal COUT is connected to port C.

ADD0: ADDER
    generic map (PROP_DELAY => 25 ns)
    port map (A(K), B(K), C(K–1), C(K), S(K));

    -- Generic map is used to pass values of generics into component instance.
    -- Positional association is used for port map.
    -- Named association is used for generic map.

M1: **component** MUX **port map** (SEL => TO_MVL(CODE),
    D0 => TO_MVL(BUS(0)), D1 => TO_MVL(BUS(1)),
    TO_BIT(Z) => CTRL);

    -- MUX is explicitly specified to be a component, even though this is not
    -- necessary.
    -- A function used in an association list represents a conversion function.
    -- A conversion function may appear in the actual part if port is either an
    -- **in** or an **inout** port; this implies that conversion function is called
    -- whose return value is then assigned to the port. Similarly, a conversion
    -- function may appear in the formal part for an **out** or an **inout** port implying
    -- that the output value is passed to a conversion function and the return
    -- value is then applied to the actual.
    -- In above example, SEL, D0 and D1 are **in** ports and Z is an **out** port.
    -- TO_MVL and TO_BIT are examples of conversion functions.

AM2: **entity** WORK.AM2910 **port map** (SEL0, SEL1, CTR_OUT);

    -- In this case, AM2 is an instance of the design entity AM2910 (the
    -- architecture, by default, is the most recently analyzed architecture body of
    -- entity AM2910).

PIP: **entity** JOE.PIPELINE (DECODER)
    generic map (NO_OF_STATE => 36)
    port map (WR, PUSH, DOUT);

    -- PIP is an instance of the design entity, architecture DECODER of
    -- entity PIPELINE, from design library JOE.

C1: **configuration** CMOS62.PARITY16CON
    generic map (TPLH => 2 ns, TPHL => 4 ns)
    port map (OPD1 => A, OPD2 => B, PAR_OUT => Z);

    -- C1 is an instance of a configuration PARITY16CON residing in design
    -- library CMOS62. The generic map and port map specify mapping to the
    -- generics and ports of the entity to which the configuration belongs.

AOI: AND_OR_GATE **port map** (D0 => ZERO, D1 => ZERO,
                                  D2 => ONE, DZ => SZ);

    -- Actuals can also be expressions (globally static). Such actuals may,
    -- however, only be associated with formals of mode **in**. In this example, D0,
    -- D1, and D2 are assumed to be of mode **in**, and ZERO and ONE are
    -- assumed to be constants.

A1: NAND2
        generic map (6.2 ns, 7.9 ns)
        port map (DR(0), DS1, RX);

    -- A component instantiation using positional association for both
    -- generic and port maps.

AX: AOI22 **port map** (DX(0), DX(1), DX(2), Z => DX_OUT, D => DX(3));

    -- Uses a mix of positional and named associations. Any positional
    -- associations must precede all named associations. In this example, the first
    -- three are positional associations, while the last two are named
    -- associations.

2.17 Concurrent assertion statement

A concurrent assertion statement executes whenever an event occurs on any signal used in the assert expression. If the assert expression is false, the specified string is printed to the output, and the specified severity level is reported to the simulator. Based on the severity level, a simulator may abort further processing, or continue processing, or issue certain diagnostic information before continuing simulation.

If the keyword **postponed** is used, then the statement will execute only at the end of a time step instead of executing in the delta in which an event occurred on a signal used in the assert expression.

Syntax

concurrent_assertion_statement

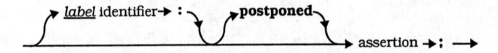

assertion

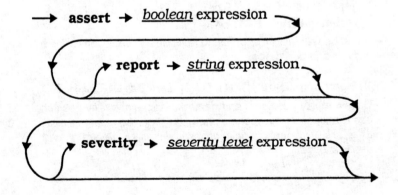

Used In

"Concurrent statement" on page 66
"Entity declaration" on page 95

Examples

assert Z(0) = '0' **and** Z(1) = '0' **and** Z(2) = '1';

    -- Executes only when an event happens on signals Z(0), Z(1) or Z(2). If the
    -- condition is false, then the default severity level of ERROR is issued to the
    -- simulator. Also the default report message "Assertion violation" is printed.

A1: **assert not** (R = '1' **and** S = '1')
    report "R and S cannot be high at the same time!"
    severity FAILURE;

    -- This statement has a label A1 and executes every time there is an event on
    -- signal R or S. When an event happens on signal R or S and if R and S are
    -- both '1', then the specified report message is printed to the output, and the
    -- severity level of FAILURE is reported to the simulator.

L1: **postponed assert** A + B = TOTAL;

    -- The **postponed** keyword ensures that the assertion check is done at the
    -- end of a time step (that is, the time step in which there is an event on
    -- A, B, or TOTAL), not on any delta time interval.
    -- For example, if A and B have events at times 14 ns, 14 ns + 1Δ, 14 ns + 2Δ,
    -- 15 ns, 15 ns + 1Δ, 16 ns, 16 ns + 1Δ, then the assertion is checked only at
    -- the end of all the deltas of times 14 ns, 15 ns, and 16 ns.

assert L'LENGTH = R'LENGTH
    report "Bounds of input operands do not match.";

    -- Executes when an event occurs on signals L or R. If condition is false, the
    -- specified message is printed, and the default severity level of ERROR is
    -- reported to the simulator.

2.18 *Concurrent procedure call statement*

A concurrent procedure call statement executes every time there is an event on a signal in its parameter list that is of mode **in** or **inout.**

If the keyword **postponed** is used, the procedure call statement will execute only at the end of the time step instead of at the delta in which an event occurs.

Syntax

concurrent_procedure_call _statement

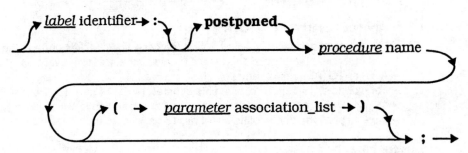

association_list

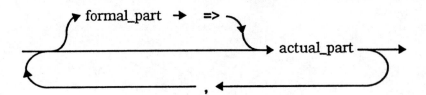

formal_part^{mod}

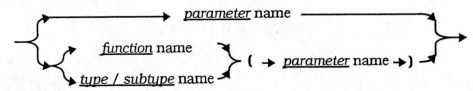

actual_part^{mod}

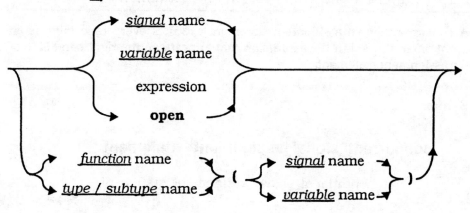

Used In

"Concurrent statement" on page 66
"Entity declaration" on page 95

Examples

P1: COUNT (ARG => S1, SUM => TOTAL);

-- TOTAL is keeping track of number of events on signal S1. Procedure
-- COUNT is called every time there is an event on signal S1. It is assumed
-- that ARG is an **in** parameter and SUM is an **out** parameter.
-- P1 is the statement label.

PRESET_CLEAR (FF => DYT, PC_VALUE => "0011");

-- This concurrent procedure call has two parameters: FF is an **inout**
-- parameter while PC_VALUE is an **in** parameter. Since a constant value
-- "0011" is being passed for PC_VALUE, any time there is an event on DYT,
-- the procedure PRESET_CLEAR is invoked.
-- This procedure call statement has no label.

P2: **postponed** SANITY_CHECK (S1, S2, S3, S4);

-- The **postponed** keyword specifies that the procedure will be called only at
-- the end of a time step instead of at delta times.

SETUP_HOLD_CHECK (D => IDATA, CLK => IBCLK);

-- A call to a procedure with two inputs, but no outputs.

2.19   *Concurrent signal assignment statement*

A concurrent signal assignment statement executes every time there is an event on any signal in the expression that appears on the right hand side of the assignment statement.

Syntax

concurrent_signal_assignment _statement

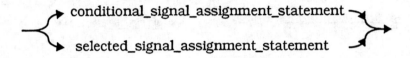

conditional_signal_assignment_statement

selected_signal_assignment_statement

Used In

"Concurrent statement" on page 66

Examples

-- The examples given below are for the simpler forms of the concurrent
-- (conditional) signal assignment that are most commonly used.

A1: Z <= (B(0) **xor** B(1)) **xor** B(2) **after** 5 ns;

-- Executes whenever there is an event on signals B(0), B(1), or B(2).
-- A1 is the statement label.

MASK <= OPCODE(7 **downto** 0) **and** DATA(16 **downto** 9);

-- Since no **after** clause is present, a delta delay is assumed by default.

RESET <= '0', '1' **after** 5 ns, '0' **after** 12 ns, '1' **after** 25 ns;

-- This statement executes once during the initialization phase of simulation,
-- which causes the four waveform elements to be deposited on the future list
-- of values for the signal RESET, and after that it is never executed again.

TX <= X"00", X"FF" **after** 2 ns, X"0F" **after** 7 ns, X"F0" **after** 11 ns;

SHIFT_R <= "000000";

-- Executes only once during the initialization phase.

```
SUM <= A xor B xor C;
    -- Delta delay is assumed by default.
```

2.20   *Concurrent statement*

A concurrent statement executes concurrently with respect to other statements, that is, it is not dependent on the textual order in which other concurrent statements are present in the design description.

Syntax

concurrent_statement

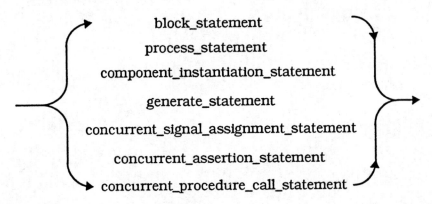

block_statement

process_statement

component_instantiation_statement

generate_statement

concurrent_signal_assignment_statement

concurrent_assertion_statement

concurrent_procedure_call_statement

Used In

"Architecture body" on page 24
"Block statement" on page 42
"Generate statement" on page 120

Examples

See respective statements.

2.21 *Conditional signal assignment statement*

The conditional signal assignment statement is a concurrent statement that behaves very similar to an if statement (an if statement is a sequential statement). In fact, a conditional signal assignment statement represents an implicit process statement with an if statement.

This statement executes whenever there is an event on any signal used in the right hand side, that is, in any of the boolean expressions or in the waveforms. The **when** conditions are checked beginning with the first, in the order in which they are written. The waveform corresponding to the first **when** boolean expression which is true is assigned to the target.

If the keyword **postponed** is used, the statement executes only at the end of the time step in which the event occurs, instead of at the delta time.

Syntax

conditional_signal_assignment_statement

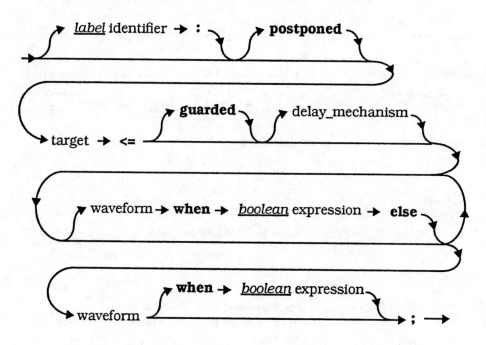

delay_mechanism

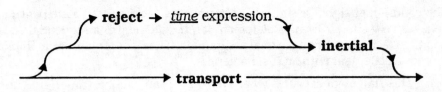

target

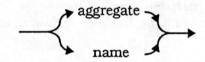

waveform^{mod}

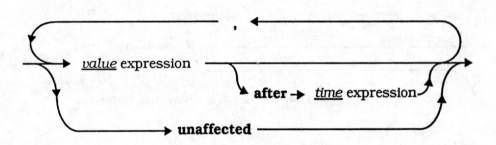

Used In

"Concurrent signal assignment statement" on page 64

Examples

G1: GATE_OUT <=   D0 **after** 5 ns **when** CTRL0 = '0' **else**
                   D1 **after** 7 ns **when** CTRL1 = '1' **else**
                   D2 **after** 9 ns;

-- G1 is the label for this conditional signal assignment statement.
-- GATE_OUT is the target signal. When an event occurs on signals D0, D1,

-- D2, CTRL0, or CTRL1, the statement executes. The first condition
-- CTRL0 = '0' is checked. If false, the second condition CTRL1 = '1' is
-- checked, and so on. If the second condition is true, then D1 is assigned to
-- GATE_OUT after 7 ns.

FSM: NXT_STATE <= S1 **when** PRESENT_STATE = S0 **else**
                  S2 **when** PRESENT_STATE = S1 **else**
                  S3 **when** PRESENT_STATE = S2 **else**
                  S0;

-- Assuming S0, S1, S2, S3 are values of an enumeration type, the above
-- statement executes when an event occurs on signal PRESENT_STATE. If
-- PRESENT_STATE has the value S2, then the value S3 is assigned to
-- signal NXT_STATE after delta delay since no delay is specified.
-- FSM is the statement label.

(COUT, SUM) <= COMPUTE_HALF_ADD (A, B, CIN) **after** ADDER_DELAY;

-- The target of this signal assignment is an aggregate. The first bit returned
-- by the function, COMPUTE_HALF_ADD, is assigned to COUT, the second
-- bit returned by the function is assigned to SUM. The following declarations
-- are assumed:

signal SUM, COUT, A, B, CIN: STD_ULOGIC;
function COMPUTE_HALF_ADD (L, R: STD_ULOGIC)
    return STD_ULOGIC_VECTOR (1 **to** 2);
constant ADDER_DELAY: TIME := 15 ns;

S1 <= A **and** B **after** 6 ns;

-- Inertial delay of 6 ns is assumed. Pulse rejection limit is same as inertial
-- delay which is 6 ns.

S2 <= **transport** B **or** C **after** 3 ns;

-- Transport delay of 3 ns is used in this case.

S3 <= **inertial** A(1) **xor** A(2) **after** 10 ns, A(2) **xor** A(3) **after** 14 ns;

-- Inertial delay and pulse rejection limit is the delay of the first waveform
-- element which is 10 ns.

CURRENT_BANK <= **reject** 5 ns **inertial** UPPER **after** 7 ns;

-- Pulse rejection limit is 5 ns while inertial delay is 7 ns.

Q <= **guarded** D **after** FF_DELAY;

-- This is a guarded signal assignment. Can occur anywhere an implicit or an
-- explicit signal called GUARD is in scope.

L2: **postponed** R_L_DATA <= R_UP_DATA (TEMP_ADDR);

-- The **postponed** keyword indicates that the signal assignment will be
-- evaluated only at the end of the time step in which an event occurred on
-- signal R_UP_DATA, instead of at the delta time in which the event
-- occurred.

BUS0 <= **unaffected when** RW = '0' **else** MEM0;

-- The keyword **unaffected** causes no change to BUS0 whenever signal RW
-- becomes '0'. It is easier to understand the semantics of this statement in
-- terms of its equivalent process statement. The keyword **unaffected** gets
-- transformed into a **null** statement in an equivalent process statement as
-- follows:

if RW = '0' **then**
    null;
else
    BUS0 <= MEM0;
end if;

-- which implies that if RW has the value '0', no action takes place and BUS0
-- remains unaffected.

2.22 *Configuration declaration*

A configuration declaration is used to describe the hierarchy of a design entity. It explicitly specifies the top-level entity-architecture pair (that is, the design entity) to which this configuration applies. The configuration declaration
also specifies the bindings of components, which appear in the architecture
body, to design entities or configurations at lower levels in the hierarchy.

A configuration declaration consists of a top-level block configuration
that specifies the binding of an entity to one of its architecture bodies. Within
this block configuration, there can be one or more component configurations
that specify the bindings of components in the architecture body to design entities or configurations. This block configuration can also contain other nested
block configurations corresponding to a block statement or a generate statement.

Syntax

configuration_declaration

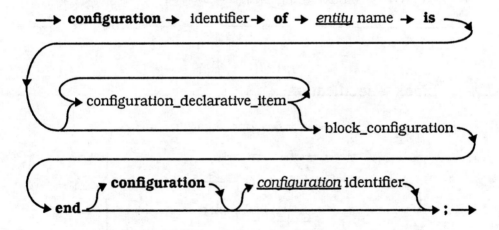

configuration_declarative_item

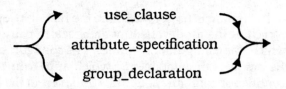

block_configuration

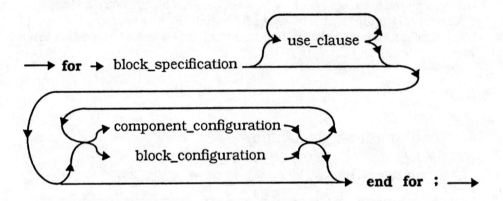

block_specification

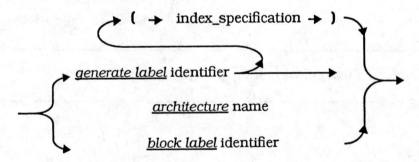

index_specification

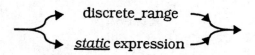

discrete_range

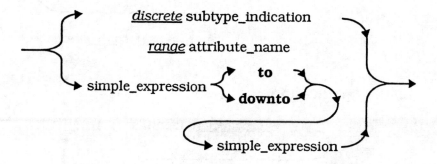

component_configuration

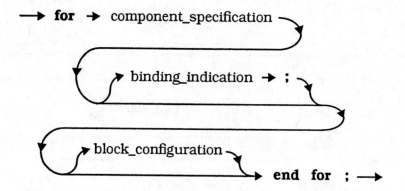

component_specification

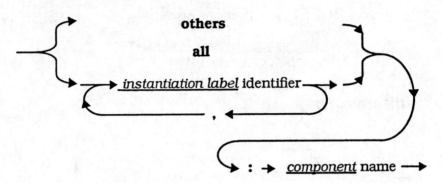

binding_indication

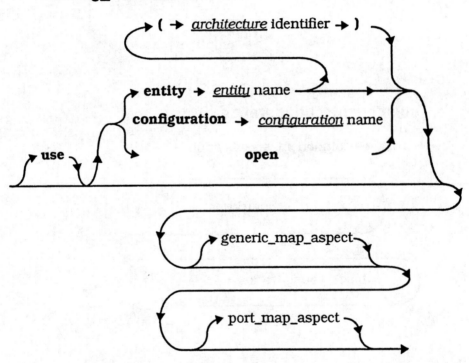

generic_map_aspect

→ **generic map (** → *generic* association_list → **)** →

port_map_aspect

→ **port map (** → *port* association_list → **)** →

association_list

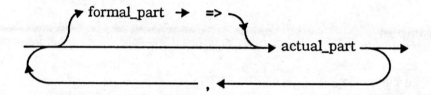

formal_part^{mod}

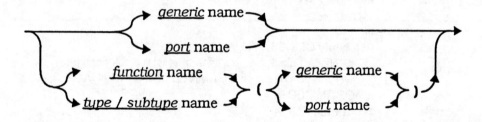

actual_part^{mod}

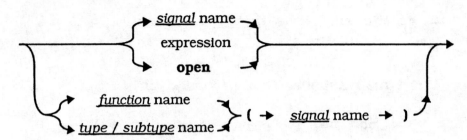

Used In

"Design file" on page 90

Examples

```
library CMOS6;
configuration RTL_CON of RTL_ENTITY is
    -- Configuration name is RTL_CON and entity name is RTL_ENTITY.
    for RTL_ARCH  -- This top-level block configuration is for the
                -- architecture RTL_ARCH belonging to entity RTL_ENTITY.
        for R1, R2: REG
            use entity WORK.UNIV_REG;
        end for;
        for all: ALU
            use entity CMOS6.SYN_ALU;
                -- Nested block configuration, that is, a block configuration
                -- for an architecture OPTIMIZED of entity SYN_ALU.
            for OPTIMIZED
                for all: XOR2
                    use entity CMOS6.XOR2; -- Default architecture is the
                            -- most recently analyzed architecture of entity XOR2.
                end for;
                for all: NAND2
                    use entity CMOS6.NAND2;
                end for;
                for all: NOR2
                    use entity CMOS6.NOR2;
                end for;
            end for;
        end for;
```

```
          for all: MUX
              use entity WORK.MUXENTITY(MUX_ARCH);
          end for;
       end for;
   end configuration RTL_CON;


library CMOS, HS_LIB;
configuration FA_CON of FULL_ADDER is
     for FA_STR        -- Top-level block configuration for architecture FA_STR
                       -- of entity FULL_ADDER residing in working library.
         for BLOCK_1      -- Block in architecture FA_STR.
             for X1, X2: XOR2
                 use entity WORK.XOR2(BEH);
             end for;
         end for;
         for BLOCK_2      -- Another block in architecture FA_STR.
             for A3: AND2
                 use entity HS_LIB.AND2HS(STR)
                     port map (HS_B => A1, HS_Z => Z, HS_A => A0);
             end for;
             for others: AND2
                 use entity WORK.AND_GATE
                     port map (A0, A1, Z);
             end for;
         end for;
         for BLOCK_3      -- Third block in architecture FA_STR.
             for all: OR2
                 use entity CMOS.OR2(C_CODE);
             end for;
         end for;
     end for;
end;           -- The keyword configuration and the configuration name are
               -- optional after the keyword end.


configuration FOUR_BIT_CON of FOUR_BIT_ADDER is
     use WORK.all;  -- A declarative item within this configuration declaration.
     for REGULAR_STR
         for GK(1)      -- Block configuration for a generate label.
             for FA: FULL_ADDER
                 use configuration FA_HA_CON;
             end for;
         end for;
```

```
            for GK(2 to 3)  -- Block configuration for two generate instances.
                for FA: FULL_ADDER
                    -- Use default bindings.
                end for;
            end for;
            for GK(0) -- Block configuration for the 0th generate instance.
                for FA: FULL_ADDER
                    use entity FULL_ADDER (CONCURRENT);
                end for;
            end for;
        end for;
end configuration FOUR_BIT_CON;

configuration OR2_CON of OR2 is
    for OR2_STR
        for N1: NOR2
            use entity WORK.NOR2(C_MODEL)
                generic map (TP_A_Z => 5 ns, TP_B_Z => 7 ns)
                port map (A, B, Z);
        end for;
        for N2: NOR2
            use entity WORK.NOR2
                generic map (6 ns, 8 ns)
                port map (A => A0, B => A1, Z => Z);
        end for;
    end for;
end configuration OR2_CON;

configuration TOP of XDOWN is
    for BEHAVIORAL
    end for;
end;
        -- A configuration declaration with no component configurations since the
        -- architecture body BEHAVIORAL has possibly no components to be bound.
```

```
configuration FA_CON_2 of WORK.FA is
    for FA_HA
        for all: HALF_ADDER
            generic map (12.2 ns, 13.6 ns);
        end for;
        for O1: OR2
            generic map (4.2 ns, 3.7 ns);
        end for;
    end for;
end;
```

-- Generic maps specified in a configuration declaration override those
-- specified earlier, possibly in a configuration specification. Bindings of
-- components to design entities or configurations have possibly also been
-- specified earlier via configuration specifications.

2.23   *Configuration specification*

A configuration specification is used to specify the binding of a component or a set of instances of a component to an entity-architecture pair (that is, to a design entity) or to a configuration. In contrast to a configuration declaration, which is a separate design unit, a configuration specification appears in the design unit that contains the component instantiation.

Syntax

configuration_specification

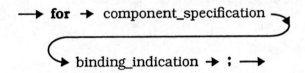

component_specification

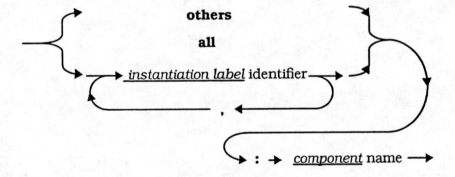

binding_indication

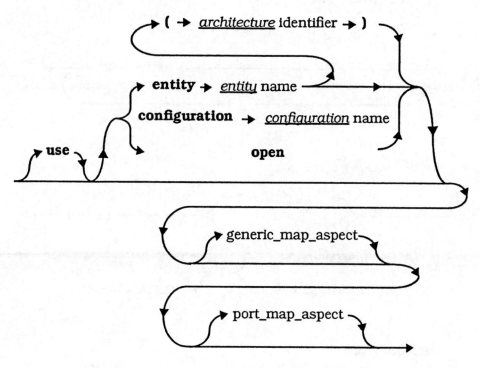

generic_map_aspect

→ **generic map (** → *generic* association_list **)** →

port_map_aspect

→ **port map (** → *port* association_list **)** →

association_list

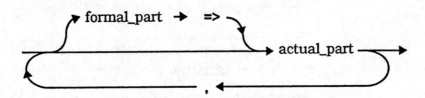

formal_part^{mod}

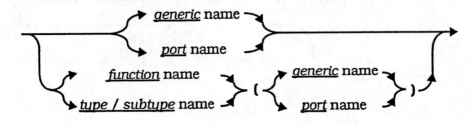

actual_part^{mod}

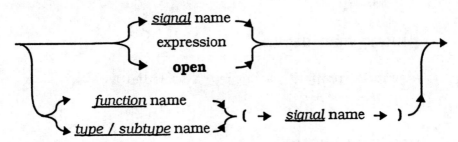

Used In

"Architecture body" on page 24
"Block statement" on page 42
"Generate statement" on page 120

Examples

for all: REG **use entity** WORK.FDSX(CALL_C_MODEL)
                    generic map (6 ns, 11 ns);
    -- For all instances of component REG, bind these to the entity FDSX and
    -- architecture body CALL_C_MODEL from the design library WORK. Values
    -- for the generics are also specified.

for C1, C2: COMPARATOR **use entity** TTL.LS74182
                    port map (A => Z0, B => Z1, Z => ZOUT);
    -- Instances C1 and C2 of component COMPARATOR are bound to the most
    -- recently analyzed architecture (since architecture name is not specified) of
    -- entity LS74182 residing in the design library TTL. Also the ports of the
    -- component (Z0, Z1, ZOUT) are bound with those of the entity (A, B, Z)
    -- since the names and possibly the order are different. Named association is
    -- used in this port map binding.

for others: NAND2 **use configuration** MYLIB.NAND2CON;
    -- Binds all remaining unbound instances of component NAND2 using a
    -- configuration named NAND2CON residing in design library MYLIB.

for X1, X2: XOR2 **use open**;
    -- Binding of instance X1 and X2 are left open, that is, bindings are
    -- deferred. May be specified later possibly in a configuration declaration.

for S1: SEQ_B **use entity** WORK.SEQUENCER;
    -- Binds instance S1 of component SEQ_B to entity SEQUENCER residing
    -- in the WORK library. The architecture body is by default the most
    -- recently analyzed architecture for that entity.

2.24 *Constant declaration*

A constant declaration declares a constant of a specified type and its value. The value of the constant is set before simulation starts, and the value cannot be changed later. However, for constants declared in subprograms, they are set each time the subprogram is invoked.

If the value of a constant is not specified in a constant declaration, then it is called a deferred constant declaration. A deferred constant declaration can only appear inside a package declaration. However a complete constant declaration, that is, a constant declaration with a value, must be present in the corresponding package body.

Syntax

constant_declaration

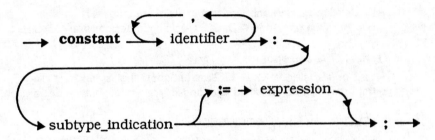

Used In

"Architecture body" on page 24
"Block statement" on page 42
"Entity declaration" on page 95
"Generate statement" on page 120
"Package body" on page 159
"Package declaration" on page 162
"Process statement" on page 172
"Subprogram body" on page 202

Examples

constant LARGEINT: INTEGER := 99999;
-- LARGEINT is a constant of type INTEGER with a value of 99999.

constant MPXVAL: BIT_VECTOR := O"57";

    -- MPXVAL is a constant bit-vector consisting of six elements. Its range,
    -- 0 **to** 5, is inferred from the value of the constant.

constant DERATING_FACTOR: REAL;

    -- This is a deferred constant since no value is specified for the constant.
    -- This constant declaration can appear only inside a package declaration.

constant DERATING_FACTOR: REAL := 0.4;

    -- Complete constant declaration for the previous deferred constant must
    -- appear in the corresponding package body.

constant S1, S2, S3: STD_LOGIC_VECTOR (7 **downto** 0) := (**others** => '1');

    -- Three array constants S1, S2, and S3, are assigned with all elements in
    -- each array set to '1'.

type MVL **is** ('U', '0', '1', 'Z');
type MVL_TBL **is array** (MVL) **of** CHARACTER;
constant TO_CHAR: MVL_TBL := ('U', '0', '1', 'Z');

    -- This is an efficient technique to convert values from one type to another
    -- type that is closely related. For example,

    variable CLA: CHAR;
    variable RPL: MVL;

    . . .
    CLA := TO_CHAR (RPL);

constant STROBE: TIME := CLK_PERIOD / 2;

    -- The value of the constant can be an expression. However, its value must
    -- be computable before simulation starts.

type MVL_1D **is array** (MVL) **of** MVL;   -- MVL is as declared above.
type MVL_2D **is array** (MVL) **of** MVL_1D;
constant TABLE_AND: MVL_2D :=
    (('U',   '0',   'U',   'U'),
    ('0',   '0',   '0',   '0'),
    ('U',   '0',   '1',   'U'),
    ('U',   '0',   'U',   'U'));

constant MEMORY: STD_ULOGIC := 'Z';

2.25   *Constrained array type declaration*

A constrained array type declaration defines an array type in which the number of elements of the array are explicitly specified.

Syntax

constrained_array_type_declaration

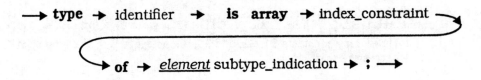

index_constraint

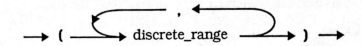

discrete_range

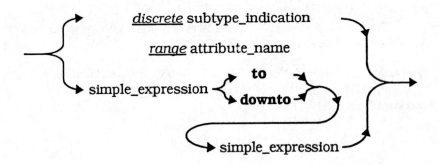

Used In

"Type declaration" on page 219

Examples

    type MVL_WORD **is array** (0 **to** 63) **of** MVL;

        -- MVL_WORD is a constrained array type of 64 elements, each element of
        -- type MVL.

    type OP_TYPE **is** (ADD, SUM, MUL, DIV);
    type TIMING **is array** (OP_TYPE) **of** TIME;

        -- TIMING is an array of four elements, each of which is indexed by a value of
        -- type OP_TYPE, and each element contains a value of type TIME.

    signal OPD1: BIT_VECTOR(15 **downto** 0);
    type NEW_VEC **is array** (OPD1'RANGE) **of** BIT;

        -- NEW_VEC is an array with 16 elements, with range 15 **downto** 0, each
        -- element is of type BIT.

    type STRING_PTR **is access** STRING (0 **to** MAX_NAME);
    type HASH_TBL **is array** (+MAX_PRIME **downto** –MAX_PRIME)
        of STRING_PTR;

        -- HASH_TBL is an array of pointers to strings.

    type MONTH **is** (JAN, FEB, MAR, APR, MAY, JUN, JUL, AUG, SEP,
        OCT, NOV, DEC);
    type SAVINGS **is array** (MONTH **range** DEC **downto** JUL) **of** REAL;

        -- SAVINGS is an array type that is indexed by values DEC downto JUL and
        -- each element stores a real value.

    type LOGIC4 **is** ('X', '0', '1', 'Z');
    type LOGIC4_1D **is array** (LOGIC4) **of** LOGIC4;

        -- LOGIC4_1D is a constrained array type indexed by the values 'X', '0', '1',
        -- and 'Z'. Each element of the array holds a value of type LOGIC4. For
        -- example,

        signal FOO: LOGIC4_1D;
        -- Then, one can write:

    FOO('0') <= 'Z';   -- Where the second element in the array of size four, FOO,
                     -- is assigned the value 'Z'.

    type BLA **is array** (NATURAL **range** 8 **to** 15) **of** LOGIC4;

        -- An example with index constraint in the form of a subtype with range
        -- constraint.

2.26 *Decimal literal*

A decimal literal represents an integer value or a real value. If written in exponent form, the exponent is always in base 10 and the default sign is + for the exponent.

Syntax

decimal_literal

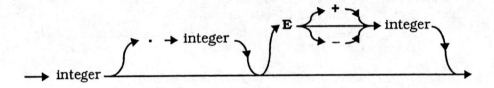

Used In

"Literal" on page 150
"Physical literal" on page 165

Examples

26
    -- An integer value

44_78_83
    -- Underscores are ignored. This is integer value: 447883

61E2
    -- An integer value in exponent form: $61*10^2 = 6100$

50.1
    -- A real value

19.563E–2
    -- A real value in exponent form: $19.563*10^{-2} = 0.19563$

6_6E+2
    -- Same as 66E2

2.27   *Design file*

A design file is the basic unit that is compiled by a VHDL analyzer, that is, a design file represents a complete legal VHDL description.

A design file consists of one or more design units. A design unit may optionally have context clauses. A context clause specifies items that are imported into the associated design unit.

An entity declaration, configuration declaration, and a package declaration are considered to be primary design units, while an architecture body and a package body are considered to be secondary design units. A secondary design unit must be compiled into the same design library in which its associated primary design unit resides.

Syntax

design_file

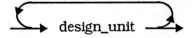

design_unit

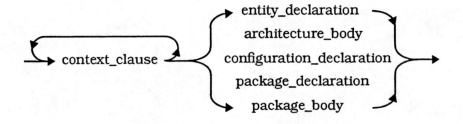

context_clause

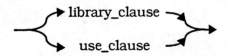

Used In

> None. This is the root construct.

Examples

-- A design file may consist of one or more of the following design units:

-- Design unit 1:
library CMOS9;   -- Library clause.
configuration FA_CON **of** FA **is**
    -- Configuration declaration example.
    -- The library clause makes the name CMOS9 visible inside the configuration
    -- declaration.
end configuration FA_CON;

-- Design unit 2:
library IEEE;
use IEEE.STD_LOGIC_1164.**all**;   -- Use clause.
    -- This use clause makes all items declared in package STD_LOGIC_1164
    -- that resides in design library IEEE to be visible in the following design unit,
    -- in this case, the package declaration, and its associated package body, if
    -- any.
package ARITH **is**
    . . .
end package ARITH;

-- Design unit 3:
library DZX, ATT;
use ATT.ATT_MVL.MVL;
use DZX.ARITH.**all**;
entity CPU **is**
    -- Library names DZX and ATT are available for use within this entity
    -- declaration and also within any of its associated architecture bodies.
    -- All items declared in package ARITH are available for use.
    -- Only the type declaration for MVL from the package ATT_MVL in
    -- library ATT is visible.
    . . .
end entity CPU;

2.28 *Digit*

A digit represents a number between 0 and 9.

Syntax

digit

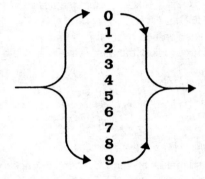

Used In

"Based literal" on page 39
"Bit string literal" on page 41
"Identifier" on page 130
"Integer" on page 139

Examples

0

9

5

2.29   *Disconnection specification*

A disconnection specification specifies the delay after which a driver of a guarded signal, which appears in a guarded signal assignment, is disconnected. The disconnection specification must appear in the declarative part enclosing the guarded signal declaration.

Syntax

disconnection_specification

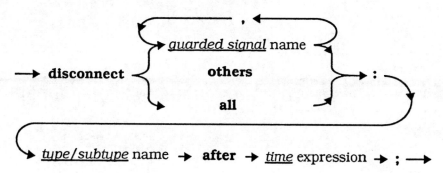

Used In

"Architecture body" on page 24
"Block statement" on page 42
"Entity declaration" on page 95
"Generate statement" on page 120
"Package declaration" on page 162

Examples

disconnect BUS_HOLD, ADDR_HOLD: STD_LOGIC_VECTOR(0 **to** 7)
                                             after 5 ns;

-- BUS_HOLD and ADDR_HOLD must be guarded signals. The drivers of
-- these signals in guarded signal assignments get disconnected 5 ns after
-- the guard becomes false.

disconnect all: RESOLVED_MVL **after** RSY_DEL;

    -- The disconnect time of RSY_DEL applies to all the drivers of guarded
    -- signals of type RESOLVED_MVL.

disconnect CPU_BUS: BIT_VECTOR (CPU_BUS'RANGE) **after** 2.6 ns;

    -- A driver of the guarded signal CPU_BUS is disconnected 2.6 ns after
    -- the guard expression becomes false.

2.30 Entity declaration

An entity declaration specifies the external interface of the design entity being modeled. It specifies the name of the entity, the number and type of ports and generics.

Ports specify the channels used by the design entity to communicate with other design entities. A port is a signal.

Generics are used to pass information into the design entity from the external environment, for example, rise and fall delays, and number of inputs.

The entity declaration may also contain certain declarations that are common to all its associated architecture bodies. In addition, it may contain certain concurrent statements that execute each time the entity is evaluated during simulation. These statements, called entity statements, must be passive, that is, they cannot update any signal value. These entity statements are useful for performing certain checks on behavior of inputs, for example, setup and hold timing violations, and data range violations.

Syntax

entity_declaration

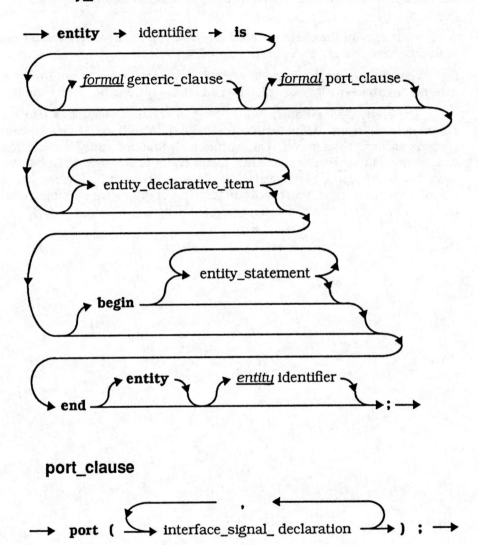

port_clause

generic_clause

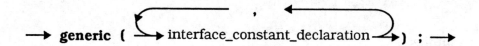

entity_declarative_item

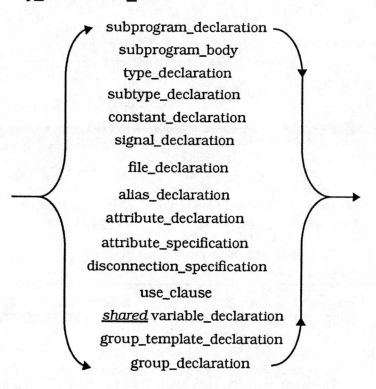

entity_statement

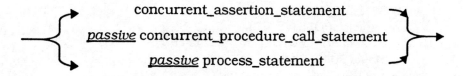

Used In

"Design file" on page 90

Examples

```
entity D_FF is                    -- Name of entity is DFF.
    port (D,
        CLK,
        CLR: in BIT;
        Q, QBAR: out BIT);    -- Port clause.
end entity D_FF;
    -- Entity declaration with only a port clause: most common form.

library NLIB; use NLIB.USEFUL.all;
entity SN7474 is
    generic (RISE_TIME, FALL_TIME: TIME); -- Generic clause.
    port (J, K, CK: in STD_LOGIC;
        Q, QBAR: out STD_LOGIC;
        VCC, VDD: in VOLTAGE);
    -- Next two are entity declarative items:
    constant SETUP_TIME: TIME := 4 ns;  -- Constant declaration.
    constant HOLD_TIME: TIME := 3 ns;   -- Constant declaration.
begin
    -- Next three statements are entity statements:
    assert CK = '0' or CK = '1'
        report "CK does not have a binary value!"; -- Conc. assertion statement.
    assert VCC <= 5 VOLTS;               -- Concurrent assertion statement.
    CHECK_TIMING_VIOLATION (J, K, CK, SETUP_TIME, HOLD_TIME);
        -- Passive concurrent procedure call.
end entity SN7474;
    -- This entity declaration has a generic clause, a port clause, entity
    -- declarations, and entity statements. The types, STD_LOGIC and
    -- VOLTAGE, and the procedure CHECK_TIMING_VIOLATION are assumed
    -- to be declared in the package USEFUL that resides in library NLIB.

entity GENERIC_AND is
    generic (M: POSITIVE := 5);
    port (A: in BIT_VECTOR(1 to M); Z: out BIT);
    attribute BUILT_IN: BOOLEAN;
end entity GENERIC_AND;
```

entity TESTBENCH **is end entity** TESTBENCH;
entity DUMMY **is end;**
    -- Two entity declarations with no ports, no generics, no declarations,
    -- and no entity statements.

entity SN7400 **is**
    port (A, B: **in** BIT_VECTOR (LOW_BIT **to** HIGH_BIT);
        C: **out** BIT_VECTOR (LOW_BIT **to** HIGH_BIT));
end;
    -- The entity name and the keyword **entity** after keyword **end** are optional.

library IEEE; **use** IEEE.STD_LOGIC_1164.**all;**
entity NOR2 **is**
    port (A, B: **in** STD_ULOGIC; Z: **out** STD_ULOGIC);
end entity NOR2;

library IEEE; **use** IEEE.STD_LOGIC_1164.**all;**
entity COUNTER **is**
    generic (N: POSITIVE := 4);
    port (ICLK: **in** STD_LOGIC; UPDOWN: **in** STD_LOGIC := '0';
        Q: **buffer** STD_LOGIC_VECTOR (1 **to** N));
end entity COUNTER;
    -- A generic has an initial value. When this entity is instantiated and if no
    -- generic value is specified, the default value of 4 is used. Similarly, a port
    -- UPDOWN has an initial value. If UPDOWN port is left unconnected when
    -- this entity is instantiated, the default value of '0' is used for this port value.

2.31   *Enumeration literal*

Enumeration literals are values of an enumeration type. These could be either character literals or identifiers that appear in an enumeration type declaration.

An enumeration literal can be used in more than one enumeration type declaration. Such a literal is said to be overloaded.

Syntax

enumeration_literal

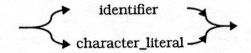

Used In

"Enumeration type declaration" on page 102
"Literal" on page 150

Examples

type BOOLEAN **is** (FALSE, TRUE);
-- Example of identifiers as enumerations literals. TRUE and FALSE are
-- enumeration literals.

type HEX **is** ('0', '1', '2', '3', '4', '5', '6', '7', '8', '9', 'A', 'B', 'C', 'D', 'E', 'F');
-- Example of character literals as enumeration literals. '0' through 'F' are
-- enumeration literals. '0' has the position number 0 and the 'F' literal has the
-- position number 15.

type LOGIC5 **is** (LOW, HIGH, RISING, FALLING, AMBIGUOUS);
-- LOW, HIGH, RISING, FALLING, AMBIGUOUS are enumeration literals
-- with LOW having the position number 0 and AMBIGUOUS having the
-- position number 4.

type LOGIC4 **is** ('X', '0', '1', 'Z');

    -- 'X', '0', '1', and 'Z' are enumeration literals. '0' is used as an enumeration
    -- literal in two type declarations, HEX and LOGIC4; '0' is said to be an
    -- overloaded literal.

type LOGIC_STATE: LOGIC5;
. . .
if LOGIC_STATE = RISING **then**

    . . .
end if;

    -- An example of an enumeration literal in a context other than a type
    -- declaration (in an expression).

2.32   *Enumeration type declaration*

An enumeration type is a type containing a set of values in the form of an ordered list.

Each of these values, which is an enumeration literal, has a position number associated with it. The leftmost literal of the type has the position number zero, the position number of any other literal is one more than that of the literal to its left. The position numbers are used in defining the ordering relations of any literal. This implies that the leftmost literal has a value less than the literal to its right, and so on.

The predefined enumeration types are CHARACTER, BIT, BOOLEAN, SEVERITY_LEVEL, FILE_OPEN_KIND, and FILE_OPEN_STATUS.

Syntax

enumeration_type_declaration

→ **type** → identifier → **is** (→ enumeration_literal →) ; →

Used In

"Type declaration" on page 219

Examples

type LOGIC **is** ('U', '0', '1', 'Z');
-- LOGIC is an enumeration type that contains four enumeration literals.
-- Each of the literals is a character. The position number of literal 'U' is 0, that
-- of '0' is 1, and so on. This also implies that 'U' < '0' < '1' < 'Z'.

type SPEED **is** (SLOW, MEDIUM, FAST);
-- SPEED is an enumeration type that contains three enumeration literals.
-- Each of the literals is an identifier. The position numbers for SLOW,
-- MEDIUM, and FAST are 0, 1, and 2 respectively.
-- Also, SLOW < MEDIUM < FAST.

type MACHINE_STATE **is** (APPLY, WAITS, READY);

type CLOCK_TYPE **is** (LOGICAL_0, LOGICAL_1, RISING, FALLING);

type MIDAS **is** (ZERO, ONE, UNKNOWN, HIGH_IMPEDANCE);

type FEELS **is** (BAD, GOOD);

2.33   Exit statement

An exit statement causes the termination of the specified loop in which it is enclosed. Execution continues with the first statement following the loop statement. If a condition is specified, the exit is executed if the condition value is true.

If no loop label is specified, the innermost loop is assumed. An exit statement can only appear inside a loop statement.

Syntax

exit_statement

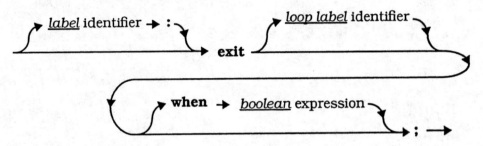

Used In

"Sequential statement" on page 188

Examples

```
N := 0; SUM := 0;
INN: loop
     SUM := SUM + N;
     E1: exit when N = 5;
          -- Since no loop label is specified, by default, the innermost
          -- loop, which is labeled INN, is exited if condition N = 5 is true.
          -- E1 is the label for the exit statement.
```

```
        N := N + 1;
end loop INN;

L1: loop
    . . .
    L2: for M in D'RANGE loop
        . . .
        if COUNT (M) = '1' then
            exit L1;
                -- Exits loop L1 when this statement is executed.
                -- Execution resumes at statement following "end loop L1;".
                -- This exit statement has no label.
        end if;
    end loop L2;
end loop L1;

-- Other forms of exit statement:

exit;
    -- Exits innermost loop unconditionally when executed.

exit LOOP_LABEL_L4;
    -- Exits specified loop when executed.

exit when COUNT > 63;
    -- Exits innermost loop only if specified condition is true, otherwise continues
    -- with statement following the exit statement.

E_LBL: exit;
    -- When executed, causes innermost loop to exit unconditionally.
    -- E_LBL is the label for the exit statement.

EX3: exit LOOP3 when Q >= (B**2 – 4*A*C) / 2;
    -- Exits specified loop LOOP3 only if condition is true, otherwise continues
    -- with next statement.
```

2.34 *Expression*

An expression is a formula for computing a value. It consists of operators and operands.

An expression is formed using one or more simple expressions (see section on "Simple expression") that are combined using logical, relational, and shift operators. The following table shows the predefined operators that can be used to form an expression.

Operator	Left operand type	Right operand type	Result type
Binary logical operators: and, or, nand, nor, xor, xnor	BIT/BOOLEAN or 1D array of BIT/BOOLEAN	Same as left and must be of same length	Same as left
Relational operators: =, /=	Any type	Same as left	BOOLEAN
<, <=, >, >=	Scalar type or discrete array type	Same as left	BOOLEAN
Shift operators: sll, srl, sla, sra, rol, ror	1D array of BIT/BOOLEAN	INTEGER	Same as left

Categories are listed in order of increasing precedence. Binary logical operators have the lowest precedence. All operators in the same category have the same precedence and are evaluated in a left to right order.

A *static expression* is an expression that can be evaluated during the elaboration phase or the compilation phase. A *locally static expression* is an expression that can be evaluated during the compilation of the design unit that contains the expression. A *globally static expression* is an expression that can be evaluated during the elaboration phase.

Syntax

expression

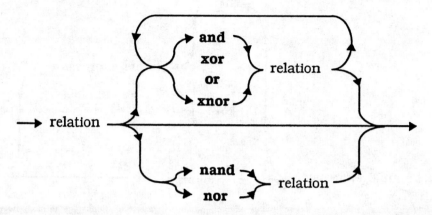

relation

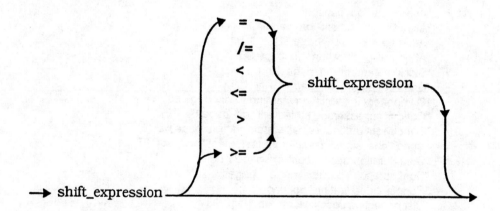

shift_expression

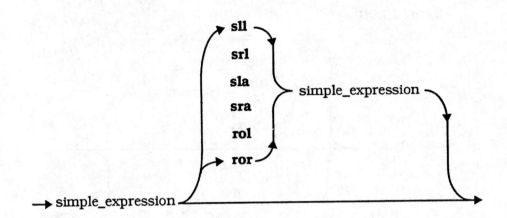

Used In

Examples

CLOCK **or** ENABLE

(DIGIT > '0') **and** (DIGIT < '9')

SIZE(A) > SIZE(B)
-- SIZE is a function call.

BUS1_REQ **and** BUS2_REQ **and** BUS3_REQ

-- No parenthesis needed. Evaluated left to right. Similar
-- for **xor, xnor,** and **or** operators.

R1 **nor** (R2 **nor** R3)

-- Parentheses required in this case since the **nor** operator is not associative.
-- Similarly parentheses are required for **nand** operator.

CTR **sll** 2

RING **ror** ROTATE_BY

(A **srl** 2) **and** (B **sll** 5)

BIT_VECTOR'("110111") **sll** 2

-- Has the value "011100". The shift left logical operator fills the vacated bits
-- with the leftmost literal of the type, which in this example is BIT'LEFT which
-- is '0'.

STD_LOGIC_VECTOR'("101111") **srl** 3

    -- Has the value "UUU101". The shift right logical operator fills the vacated
    -- bits with the leftmost literal of the type, which in this case is
    -- STD_LOGIC'LEFT, which is 'U'.
    -- Assumed that a user-defined overloaded function for the **srl** operator
    -- exists.

STD_LOGIC_VECTOR'("01XXU0Z") **sla** 4

    -- Has value "U0ZZZZZ". The shift left arithmetic operator fills the vacated
    -- bits with rightmost literal of the left operand, which in this case is 'Z'.
    -- Assumed that a user-defined overloaded function for the **sla** operator
    -- exists.

BIT_VECTOR'("1011000") **sra** 2

    -- Has the value "1110110". The shift right arithmetic operator fills the vacated
    -- bits with the leftmost literal of the left operand, which in this case is '1'.

"101111" **rol** 4

    -- Has the value "111011". The rotate left logical operator rotates the bits from
    -- left to right the specified number of times.

"000111" **ror** 2

    -- Has the value "110001". The rotate right logical operator rotates the bits
    -- right to left the specified number of times.

2.35   *File declaration*

A file declaration declares a file and associates it with a file in the host environment with the specified path name. It also specifies whether the file is a read-only or a write-only file. The file belongs to the specified file type, that is, values in the file are of the element type specified in the file type declaration.

If the file open information is specified, then the file is opened during the elaboration phase with an implicit call to procedure FILE_OPEN. If the file open information is specified for a file declared within a subprogram, the file is opened when the subprogram is called.

Syntax

file_declaration

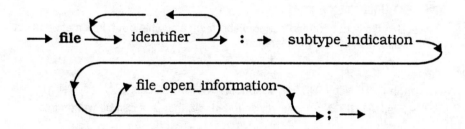

file_open_information

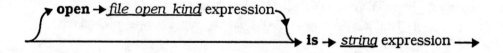

Used In

"Package declaration" on page 162
"Process statement" on page 172
"Subprogram body" on page 202

Examples

type BV_FTYPE **is file of** BIT_VECTOR;   -- File type declaration.
file VEC_FILE: BV_FTYPE **is** "/usr/home/jb/vec.in";

    -- VEC_FILE is a file. It stores bit_vectors only. It is an input file since file
    -- open kind is not specified and the default is READ_MODE. The file is
    -- associated with the physical file /usr/home/jb/vec.in, that is interpreted as a
    -- path name to a file in the host environment. This file is opened during
    -- elaboration since file open information is present.

file DISPLAY_FILE: DISPLAY_RECORDS_FTYPE
    open WRITE_MODE **is** "A:RX.HDR";

    -- DISPLAY_FILE is a file that is written to.

file TEXT_FILE: TEXT;

    -- This file is not opened during elaboration since file open information is not
    -- specified in the file declaration.

file RESULT: TEXT
    open APPEND_MODE **is** "/usr/home/XWIN/result.out";

    -- File RESULT is opened in append mode, that is, it can be written into.

file VEC1, VEC2: TEXT;

    -- Two files, VEC1 and VEC2, of file type TEXT are declared. These files
    -- have to be explicitly opened since no file open information is present.

file IN_FILE: TEXT **open** READ_MODE **is** "post.dat";

    -- IN_FILE is a file of type TEXT. It will be opened in read mode during
    -- elaboration.

file OUT_FILE: TEXT **open** WRITE_MODE **is** "fir3_out.data";

    -- OUT_FILE is a write-only file.

2.36   *File type declaration*

A file type declaration declares a file type, that is, it specifies the collection of values that a file might have.

Every file type declaration implicitly defines operations on files of the specified file type. These are the procedures FILE_OPEN, FILE_CLOSE, READ, and WRITE, and the function ENDFILE.

Syntax

file_type_declaration

$\longrightarrow$ **type** $\rightarrow$ identifier $\rightarrow$ **is file of** $\rightarrow$ *type/subtype* name $\rightarrow$; $\longrightarrow$

Used In

"Type declaration" on page 219

Examples

```
type INDEX is range 0 to 15;
    -- Integer type declaration.
type INT_FTYPE is file of INDEX;
    -- INT_FTYPE is a file type of INDEX values. The file type declaration
    -- implicitly declares the following operations:

procedure FILE_OPEN (file F: INT_FTYPE;
    EXTERNAL_NAME: in STRING;
    OPEN_KIND: in FILE_OPEN_KIND := READ_MODE);
procedure FILE_OPEN (STATUS: out FILE_OPEN_STATUS;
    file F: INT_FTYPE; EXTERNAL_NAME: in STRING;
    OPEN_KIND: in FILE_OPEN_KIND := READ_MODE);
    -- The above two procedures open the file specified in EXTERNAL_NAME in
    -- the specified mode OPEN_KIND and returns the file pointer in F. The
    -- second FILE_OPEN procedure returns the status of the procedure, if any,
    -- in STATUS.
procedure FILE_CLOSE (file F: INT_FTYPE);
```

-- Closes the specified file.
procedure READ (**file** F: INT_FTYPE; VALUE: **out** INDEX);
-- Reads a VALUE of type INDEX from file F.
procedure WRITE (**file** F: INT_FTYPE; VALUE: **in** INDEX);
-- Writes a VALUE of type INDEX to file F.
function ENDFILE (**file** F: INT_FTYPE) **return** BOOLEAN;
-- Returns true if end-of-file of file F has been reached.

type MVL_VECTOR **is array** (NATURAL **range** <>) **of** MVL;
type MVLV_FTYPE **is file of** MVL_VECTOR;

-- MVLV_FTYPE is a file type containing values of type MVL_VECTOR.
-- Since MVLV_FTYPE is a file type containing unconstrained array values,
-- the READ operation is implicitly declared in the following form:

procedure READ (**file** F: MVLV_FTYPE; VALUE: **out** MVL_VECTOR;
     LENGTH: **out** NATURAL);

type DISPLAY_RECORDS_FTYPE **is file of** DISPLAY_RECORD;
-- DISPLAY_RECORDS_FTYPE is a file type consisting of DISPLAY records.

type BV_FTYPE **is file of** BIT_VECTOR;
-- BV_FTYPE is a file type containing BIT_VECTOR values.

type BOOL_FTYPE **is file of** BOOLEAN;
-- BOOL_FTYPE is a file type containing BOOLEAN values.

type TEXT **is file of** STRING;
-- TEXT is a file type containing strings, that is, a sequence of characters.

2.37   *Floating type declaration*

A floating type declaration declares a type containing real numbers with a specified range. The range could be either an ascending or a descending range, or it could be expressed using one of the predefined range attributes ('RANGE, 'REVERSE_RANGE).

The only predefined floating type is REAL.

Syntax

floating_type_declaration

$\longrightarrow$ **type** $\rightarrow$ identifier $\rightarrow$ **is   range** $\rightarrow$ *real* range $\rightarrow$; $\longrightarrow$

range

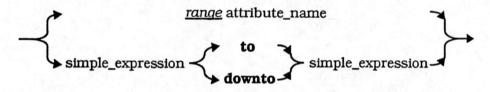

Used In

"Type declaration" on page 219

Examples

type CURRENT **is range** 1.0 **to** 15.0;
    -- CURRENT is a floating point type with values in the range 1.0 through 15.0

type NORMALIZED **is range** +1.0 **downto** –1.0;
    -- Note that negative bounds are also allowed.

type LOCAL_CURRENT **is range** CURRENT'RANGE;
    -- Type LOCAL_CURRENT also has the range 1.0 to 15.0. This type is
    -- distinct from type CURRENT declared above.

type NULL_RANGE **is range** +1.0 **downto** +2.0;
    -- This range is a null range since it does not contain any values,
    -- since 2.0 > 1.0.
    -- Consequently, the NULL_RANGE type also does not contain any values.

2.38   Function call

A function call is a call to a function, and it appears as part of an expression. A function always returns a value.

If a formal parameter of a function is a signal or a file, then the actual passed in must also be a signal or a file, respectively. If the formal parameter is a constant, then the actual can be an expression. If no object class is specified for a formal parameter, constant is assumed.

For a formal parameter that is a constant, the value of the actual is copied into the formal parameter. For a formal parameter that is a signal or a file, its reference is passed into the corresponding formal parameter.

Syntax

function_call

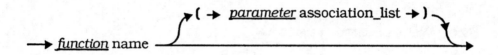

association_list

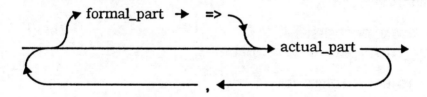

formal_partmod

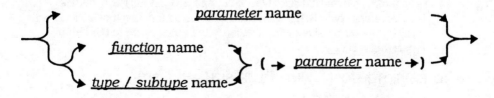

actual_part^{mod}

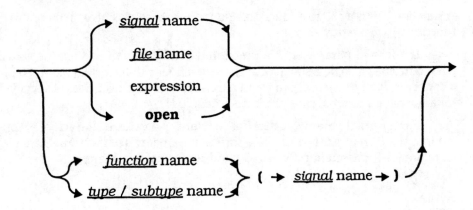

Used In

"Attribute name" on page 32
"Indexed name" on page 137
"Selected name" on page 182
"Simple expression" on page 195
"Slice name" on page 199

Examples

TOTAL_PROCESSES
    -- A function call with no parameters.

TO_BITVECTOR(56, 10)
    -- Positional association is used to pass parameters, 56 is the value of the
    -- first parameter, 10 is the value of the second parameter.

COUNT_EVENT (SIG_EVENT => TO_BIT(RST))
    -- Assuming signal RST is of type MVL, and SIG_EVENT is a signal
    -- parameter of type BIT. TO_BIT is a conversion function that is called with
    -- RST; the return value of the function is then passed on to the formal
    -- parameter SIG_EVENT.

TO_BITVECTOR (OPD => INT_VALUE, SIZE => INT_BITS);
    -- Named association used.

ADD (OPD1 => 2, OPD2 => 5, OPD3 => **open**)

    -- The **open** keyword indicates no value being passed. In this case, the third
    -- parameter of the function must explicitly have a default value specified in its
    -- parameter declaration, which is used as the value of the third parameter
    -- during function evaluation. For example, the function declaration might
    -- appear as:

    function ADD (OPD1: INTEGER; OPD2: INTEGER;
            OPD3: INTEGER := 0) **return** INTEGER;

    -- A function call of form ADD (OPD1 => 12, OPD2 => 50) will use the default
    -- value of 0 for OPD3.

TO_STDLOGICVECTOR (COUNT, SIG'LENGTH)

    -- An integer COUNT and the number of bits in the bit-vector SIG, are passed
    -- in as values to the function using positional association.

COMPUTE_DIFFERENTIAL (A * B, A + B)

    -- Function parameters can also be expressions.

READ_LINE (F => BIT_PATTERN_FILE)

    -- Function returns the next line read from the specified file.

2.39   *Generate statement*

A generate statement causes elaboration time repetition of or conditional selection from a set of concurrent statements.

There are two forms of the generate statement. The for-generate scheme causes a set of concurrent statements to be replicated for each value of the generate parameter. The value of the generate parameter is substituted in each copy of the concurrent statement. The range of the generate parameter must be determinable at elaboration time.

The second form is the if-generate scheme that causes conditional selection from a set of concurrent statements. The value of the condition must be determinable at elaboration time since a generate statement is expanded during elaboration time.

Syntax

generate_statement

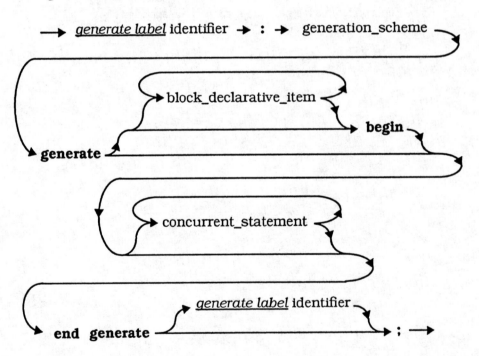

generation_scheme

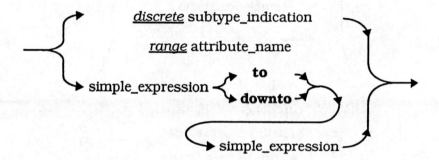

discrete_range

simple_expression

to
downto

simple_expression

discrete subtype_indication

range attribute_name

block_declarative_item

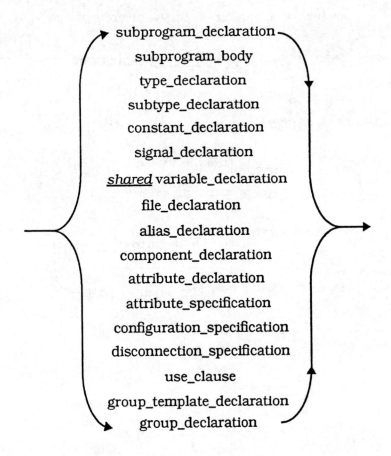

subprogram_declaration
subprogram_body
type_declaration
subtype_declaration
constant_declaration
signal_declaration
shared variable_declaration
file_declaration
alias_declaration
component_declaration
attribute_declaration
attribute_specification
configuration_specification
disconnection_specification
use_clause
group_template_declaration
group_declaration

Used In

"Concurrent statement" on page 66

Examples

G_ONE: **for** K **in** 4 **downto** 1 **generate**
   Z(K) <= A(K) **and** B(K);
end generate G_ONE;
     -- G_ONE is the generate statement label. This is an example of a
     -- for-generate scheme, where K is the generate parameter.
     -- Elaboration causes it to expand to the following four block statements:

```
        G_ONE: block
            constant K: INTEGER := 4;
        begin
            Z(K) <= A(K) and B(K);
        end block G_ONE;

        G_ONE: block
            constant K: INTEGER := 3;
        begin
            Z(K) <= A(K) and B(K);
        end block G_ONE;

        G_ONE: block
            constant K: INTEGER := 2;
        begin
            Z(K) <= A(K) and B(K);
        end block G_ONE;

        G_ONE: block
            constant K: INTEGER := 1;
        begin
            Z(K) <= A(K) and B(K);
        end block G_ONE;

GEN1: if TEST_MODE generate
    CD: entity WORK.CLOCK_DIVIDER port map (CK, RST, SLOW_CK);
end generate;
        -- This is an example of an if-generate scheme. During elaboration, if
        -- TEST_MODE has the value TRUE, then the single component instantiation
        -- statement is included in the design unit in which it appears, else it is not
        -- included.

MULTI_BIT: for M in A'RANGE generate
    signal TEMP1: BIT;
begin
    TEMP1 <= A(M) xor B(M) after 4 ns;
    SUM(M) <= TEMP1 xor C(M–1) after 5 ns;
    CL: entity LIBX.CARRY_LOGIC port map (C(M–1), A(M), B(M), C(M));
end generate MULTI_BIT;
        -- This generate statement has a block declarative item: this declaration
        -- is local to the generate statement. A'RANGE must be known at
        -- elaboration time.
```

2.40   *Graphic character*

Each graphic character corresponds to a unique code of the ISO 8-bit coded character set (ISO 8859-1:1987(E)) and is represented by a graphical symbol. The *most commonly used* graphic characters in English are:

- upper case letters: A B C D E F G H I J K L M N O P Q R S T U V W X Y Z

- lower case letters: a b c d e f g h i j k l m n o p q r s t u v w x y z

- digits: 0 1 2 3 4 5 6 7 8 9

- space character

- special characters: " # & ' ( ) * + , – . / : ; < = > _ | ! $ % @ ? [\ ] ^ ` { } ~

2.41   Group declaration

A group is declared using a group declaration. The group declaration specifies the items that form the group.

Syntax

group_declaration

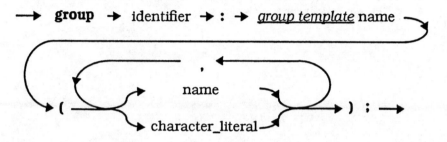

Used In

"Architecture body" on page 24
"Block statement" on page 42
"Configuration declaration" on page 71
"Entity declaration" on page 95
"Generate statement" on page 120
"Package body" on page 159
"Package declaration" on page 162
"Process statement" on page 172
"Subprogram body" on page 202

Examples

group M1: MARK (ALU3, COMP05, MUX32);

-- M1 is the name of the group that is an instance of group template MARK
-- and has labels ALU3, COMP05, and MUX32 as members.

group K1: KEEP (RST, RDY, SDRD);

-- K1 is the name of the group, of group template KEEP and has the
-- signals RST, RDY, and SDRD as members.

group COMB1: PORT2CLOCK (D, CLOCK);
group COMB2: PORT2CLOCK (PC, CLOCK);
group COMB3: PORT2CLOCK (Q, CLOCK);
group COMB4: PORT2CLOCK (QBAR, CLOCK);
    -- Above are four groups of the same group template.

group B1: BINDING (ALU3, GLIB.OR4);
    -- B1 is a group that contains a label and an entity name pair.

group A_GRP: EQUIVALENT ('A', 'a');
    -- This group contains two elements.

2.42   *Group template declaration*

A group is a set of named items based on the group template. A group template specifies the class of named items, for example, constant, label, or signal, that may appear in a group. A group template is declared using a group template declaration.

      The <> (box) symbol is used to specify zero or more of the specified entity class, and if it appears, it must be the last one.

Syntax

group_template_declaration

entity_class

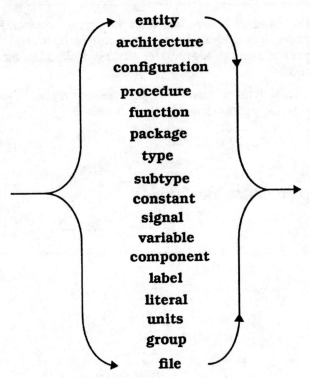

Used In

"Architecture body" on page 24
"Block statement" on page 42
"Entity declaration" on page 95
"Generate statement" on page 120
"Package body" on page 159
"Package declaration" on page 162
"Process statement" on page 172
"Subprogram body" on page 202

Examples

group MARK **is (label** <>**);**
    -- MARK is a group template that consists of zero or more labels.

group KEEP **is** (**signal** <>);
    -- KEEP is a group template that consists of zero or more signals.

group PORT2CLOCK **is** (**signal, signal**);
    -- PORT2CLOCK is a group template that contains two signals, presumably
    -- the first is a data port and the second is a clock signal.

group BINDING **is** (**label, entity**);
    -- BINDING is a group template that consists of a label and an entity name
    -- pair.

group EQUIVALENT **is** (**literal** <>);
    -- Any number of literals, including zero, can be in this group.

group ALLOCATE **is** (**signal, label** <>);
    -- This group template consists of one signal name and zero or more labels.

2.43 *Identifier*

There are two kinds of identifiers, a basic identifier and an extended identifier.

A basic identifier must begin with a letter. Remaining characters can be letters or digits or underlines. Two successive underlines are not allowed. Also, the last character may not be an underline. A letter can be either an uppercase or a lowercase character. Letters used in a basic identifier are case-insensitive. Therefore, using an uppercase character is equivalent to using a lowercase character.

An extended identifier is a sequence of graphic characters written between two backslashes ('\'). Extended identifiers are case-sensitive.

Syntax

identifier

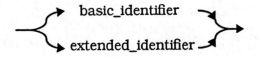

basic_identifier

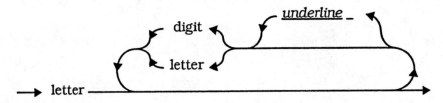

extended_identifier

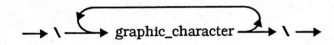

letter

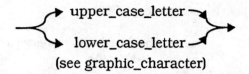

(see graphic_character)

Used In

Nearly every construct.

Examples

-- Examples of basic identifiers:

CMOS

DONE

TWOS_COMPLEMENT

RIPPLE
Ripple   -- Same identifier as RIPPLE.

COUNT3_1S

-- Examples of extended identifiers:

\.RATIO\

\23.67\

\TTL.23\

\WHILE\   -- Distinct from the keyword **while**.

_L1\

_$.!"\

\DONE\
\done\   -- This is a different identifier from \DONE\, which is different
                -- from a basic identifier DONE.

\CUT\\4\\PIECES\   -- Use two backslashes to represent a backslash inside
                      -- an extended identifier.

2.44   If statement

An if statement checks a sequence of conditions. It executes the sequence of statements that follow the first condition that evaluates to true.

Syntax

if_statement

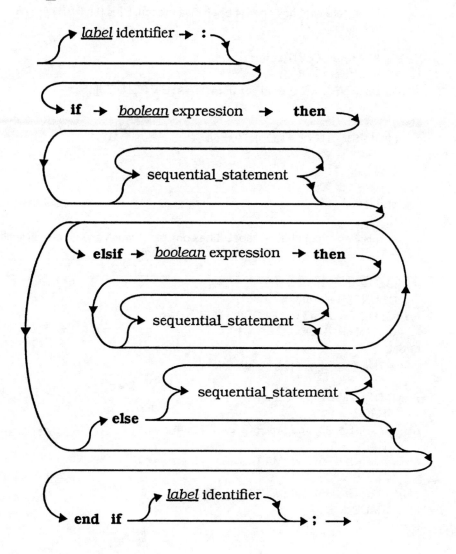

Used In

"Sequential statement" on page 188

Examples

```
L1: if LOAD = '0' then
    FF1 <= "0100";
end if L1;
```
    -- This if statement has no elsif or an else branch. L1 is the label of the if
    -- statement.

```
if FSM_STATE = COUNT0 then
    RPY := '1';
elsif (FSM_STATE = COUNT1) or (FSM_STATE = COUNT7) then
    RPY := 'Z';
else
    L2: if FSM_STATE = COUNT9 then
        RPY := '0';
    else
        RPY := 'X';
    end if L2;
end if;
```
    -- This if statement has no label. It has one elsif branch and an else branch.
    -- The else branch contains another if statement.

```
if RESET = '0' then
    COUNTER <= (others => '0');
elsif CLOCK = '0' and not CLOCK'STABLE then
    COUNTER <= COUNTER + 1;
end if;
```
    -- This if statement has no else branch.

```
BANDING: if MARKS <= 20 then
    GRADE := 'F';
elsif (MARKS > 20) and (MARKS <= 50) then
    GRADE := 'D';
elsif (MARKS > 50) and (MARKS <= 65) then
    GRADE := 'C';
elsif (MARKS > 65) and (MARKS <= 80) then
    GRADE := 'B';
else
```

```
        GRADE := 'A';
end if;
```

        -- An if statement with many elsif branches, one else branch, and a label for
        -- the if statement specified.

```
if PRESET = '1' then
        MASK <= PRE_DATA;
end if;
```

        -- An if statement with no elsif or an else branch.

2.45   Incomplete type declaration

An incomplete type declaration declares the name of a type without specifying the set of values that form the type. However, a complete type declaration must appear later in the same declarative part in which the incomplete type declaration appears. It is often used when recursive types or mutually dependent types are being described.

Syntax

incomplete_type_declaration

$\longrightarrow$ **type** $\rightarrow$ identifier $\rightarrow$; $\longrightarrow$

Used In

"Type declaration" on page 219

Examples

```
type DFG_NODE;                        -- Incomplete type declaration.
    -- The values in the type DFG_NODE are not defined yet.

type DFG_PTR is access DFG_NODE;
    -- DFG_PTR is a pointer to DFG_NODE.

type DFG_NODE is
    record
        OP: OP_CODE;
        SUCC_LIST: DFG_PTR;
        PRED_LIST: DFG_PTR;
    end record DFG_NODE;
    -- This is the complete type declaration for DFG_NODE.
```

2.46 *Indexed name*

An indexed name is a name that represents a particular element of an array. It consists of a name (an array name) or a function call (that returns an array) followed by a list of index values, each representing an index of a dimension of the array.

Syntax

indexed_name

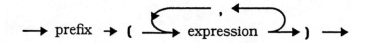

prefix

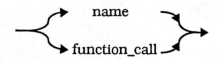

Used In

"Name" on page 155

Examples

COUNTER(0)
-- 0th bit of COUNTER.

SUM(L+1)
-- (L+1) bit of SUM.

MEMORY (K–1, J+1)
-- MEMORY is a two-dimensional array. This index name refers to the
-- element at (K–1) position in the first dimension, and (J+1) position in
-- the second dimension.

RAM (0, 127)

"∗" (A, B) (OVFL_BIT)
    -- Refers to the OVFL_BIT of the return value of the "∗" function.

AND_TABLE ('U')('Z')
    -- AND_TABLE is an array of arrays. 'U' is the index into AND_TABLE and 'Z'
    -- is the index into the resulting array.

RAM (RAM_ADDR)

2.47 *Integer*

An integer value represents a number made up of digits and underlines. Underlines are meant for documentation and are ignored.

Syntax

integer

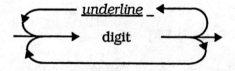

Used In

"Based literal" on page 39
"Decimal literal" on page 88

Examples

56_42
   -- Same as integer 5642.

255

1_024
   -- Same as integer 1024.

1_000_000
   -- Same as 1000000.

2.48   Integer type declaration

An integer type declares a type containing a set of integers within a given range. The range can be either ascending or descending, or it can be derived from any of the predefined range attributes.

The predefined integer type is INTEGER, and the predefined integer subtypes are POSITIVE and NATURAL.

Syntax

integer_type_declaration

⟶ **type** → identifier → **is range** → *integer* range → **;** ⟶

range

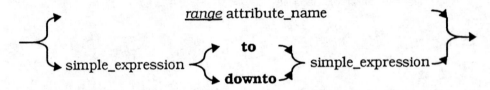

Used In

"Type declaration" on page 219

Examples

type PORT_SIZE **is range** 1 **to** 20;
    -- PORT_SIZE is an integer type with values in the range 1 through 20.

type BYTE **is range** 7 **downto** 0;

type WORD **is range** BYTE'REVERSE_RANGE;
    -- WORD is an integer type whose range is same as that of type BYTE but in
    -- the opposite direction, that is, it is of range "0 **to** 7".

type UNSIGNED **is range** 0 **to** (2∗∗16 − 1);

    -- A range value may also be an expression.

type CLIPPING **is range** −31 **to** +31;

    -- Negative values are also allowed in a range.

type BOUND_1 **is range** −5 **downto** −10;

type FILLER **is range** 5 **to** 1;

    -- The range is a null range since it does not represent any values.
    -- Consequently, the integer type FILLER also does not contain any values.

2.49   Interface constant declaration

An interface constant declaration declares a constant. A constant declared using an interface declaration is always of mode **in**, even if the mode is not explicitly specified. In certain cases, the class **constant** is assumed by default and need not be explicitly specified; these are for function parameters and for procedure parameters of mode **in**. If a static expression is specified in the declaration, then the constant assumes this value if there is no corresponding actual passed in or if **open** is specified in the interface.

Syntax

interface_constant_declaration

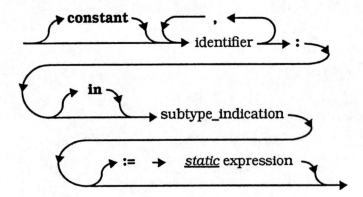

Used In

"Block statement" on page 42
"Component declaration" on page 53
"Entity declaration" on page 95
"Subprogram body" on page 202
"Subprogram declaration" on page 208

Examples

function ES_RISING (CLOCK: BIT) **return** BOOLEAN;

    -- CLOCK is a constant, since all function parameters are of constant
    -- class unless explicitly specified otherwise. It is also of mode **in**, by default.

procedure PRESET_CLEAR (**signal** FF: **inout** STD_LOGIC;
                                        PC_VALUE: **in** STD_LOGIC := '1');

    -- PC_VALUE is a constant since procedure parameters of mode **in**
    -- are, by default, constants. A default value of '1' is used if the calling
    -- program does not pass in a value for PC_VALUE or if it specifies **open.**

function MAX_DLY (**constant** RISE_DLY, FALL_DLY: **in** TIME) **return** TIME;

    -- The mode and class for the two function parameters are explicitly specified.

2.50   *Interface file declaration*

An interface file declaration declares a file. The type specified by the subtype indication must be a file type.

Syntax

interface_file_declaration

$$\longrightarrow \textbf{file} \longrightarrow \text{identifier} \longrightarrow : \longrightarrow \text{subtype\_indication} \longrightarrow$$

Used In

"Subprogram body" on page 202
"Subprogram declaration" on page 208

Examples

procedure DUMP_VECTORS (**file** LSL_FILE: X01Z_FTYPE;
                                    DATA: **in** X01Z_VECTOR);
   -- LSL_FILE is a file of file type X01Z_FTYPE.

procedure READ_INTEGERS (**file** PAT_FILE: INTEGER_FTYPE;
                                    R_VALUE: **out** INTEGER);
   -- PAT_FILE is a file of file type INTEGER_FTYPE.

2.51   *Interface signal declaration*

An interface signal declaration declares a signal. If no mode is specified, the default is mode **in**. If no class is specified, in certain cases, class **signal** is assumed. These are for ports declared in entity declarations, block statements, and in component declarations.

A static expression, if specified, is the value used by the signal if no actual signal is specified for the interface signal or if **open** is specified. It is illegal to specify a static expression for an interface signal declared in a subprogram.

Syntax

interface_signal_declaration

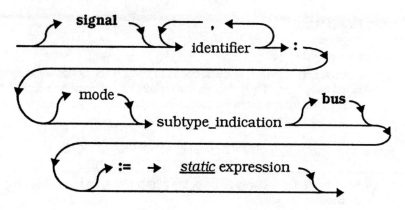

mode

Used In

"Block statement" on page 42
"Component declaration" on page 53
"Entity declaration" on page 95
"Subprogram body" on page 202
"Subprogram declaration" on page 208

Examples

component PARITY **is**
    port (A0, A1, A2, A3: STD_LOGIC; Z: **out** STD_LOGIC);
end component PARITY;
    -- All component ports are signals. Therefore, the signal class need not be
    -- explicitly specified. Since mode of signals A0, A1, A2, A3 are not specified,
    -- by default, they are assumed to be of mode **in**.

entity NAND **is**
    port (**signal** A, B: **in** BIT := '0'; Z: **out** BIT);
end entity NAND;
    -- Ports A and B are explicitly declared to be of class signal, even though this
    -- is not necessary. They also have a default value specified which is the
    -- value used for the ports if either of the input ports is left unconnected. Z is
    -- also a signal since all ports are signals.

procedure PRESET_CLEAR (**signal** FF: **inout** STD_ULOGIC;
                            PC_VAL: STD_ULOGIC);
    -- FF is explicitly declared to be of the signal class, and its mode is specified
    -- as **inout**.

library IEEE; **use** IEEE.STD_LOGIC_1164.**all**;
entity MUX **is**
    port (A0, A1, A2, A3: **in** STD_ULOGIC; Z: **out** STD_LOGIC **bus**);
end MUX;
    -- The signal Z is of **bus** signal kind.

2.52   *Interface variable declaration*

An interface variable declaration declares a variable. If no mode is specified, the default is mode **in**. If no class is specified, then for certain cases, class **variable** is assumed. These are for procedure parameters of mode **out** and **inout**.

A static expression, if present, is used as the parameter value if no actual variable is specified for this interface parameter or if **open** is specified. It is illegal to specify a static expression for an interface variable in a subprogram declaration that is of mode other than **in**.

Syntax

interface_variable_declaration

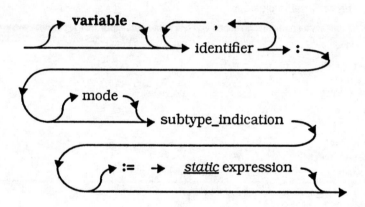

mode^mod

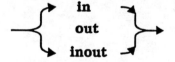

Used In

"Subprogram body" on page 202
"Subprogram declaration" on page 208

Examples

procedure COUNT_0S (SUM: **out** POSITIVE; **variable** INCR_BY: POSITIVE);

-- SUM is an output parameter with the class not explicitly specified; therefore
-- by default, it is a variable. INCR_BY has been explicitly specified to be of
-- variable class. Its mode is **in**, by default.

procedure CHECK_PATTERN (**variable** A0, A1, A2, A3: **in** BIT := '1'; Z: **out** BIT);

-- A0, A1, A2, A3 are variables with a default value of '1'. If an actual value is
-- not specified for any of these parameters, then a value of '1' is used as its
-- default value.

procedure ASYNC_RESET (ASYNC_VAL: STD_LOGIC_VECTOR;
                            variable TARGET_VAR: **out** STD_LOGIC_VECTOR);

-- TARGET_VAR is an output parameter that has been explicitly specified to
-- be of variable class.

2.53   *Library clause*

A library clause declares logical names of design libraries that may be referenced in an associated design unit.

Syntax

library_clause

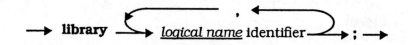

Used In

"Design file" on page 90

Examples

library ALIB, BLIB, CLIB;
    -- Declares three design libraries. These can now be referenced in the
    -- design unit with which this clause is associated.

library IEEE;
library DZX;
configuration FA_CON of FULL_ADDER is
    -- The design library names, IEEE and DZX, can now be referenced within
    -- this design unit.
    . . .
end configuration FA_CON;

library STD, WORK;
    -- This library clause is implicitly declared in all design units.

2.54   *Literal*

A literal is a value of a type. That is, the values used to form a type are called literals.

A decimal literal or a based literal is also called an abstract literal.

A numeric literal is a decimal literal, a based literal, or a physical literal.

Syntax

literal

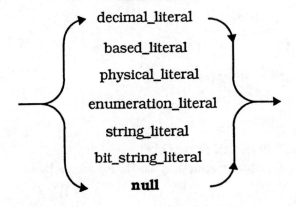

Used In

"Simple expression" on page 195

Examples

```
3.11        56_26
    -- Decimal literals.

16#FF0F#    8#372#
    -- Based literals.

3 V     5 na    ms
    -- Physical literals.
```

TRUE   READY   '0'
 -- Enumeration literals.

"ERROR"   "EXIT ON 1"
 -- String literals.

X"F0_0A"   B"1001"
 -- Bit-string literals.

null
 -- Value of an access type.

2.55   Loop statement

A loop statement causes repetitive execution of a set of sequential statements. There are three forms of the loop statement:

1. for-loop ... end loop
2. while-loop ... end loop
3. loop ... end loop

Syntax

loop_statement

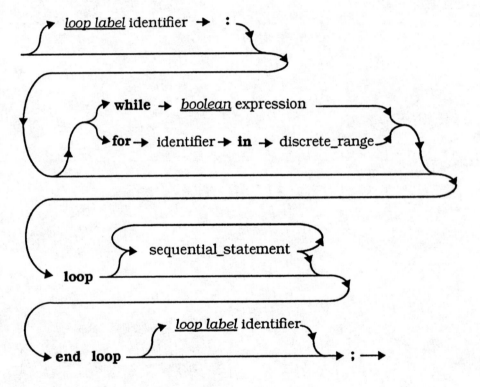

discrete_range

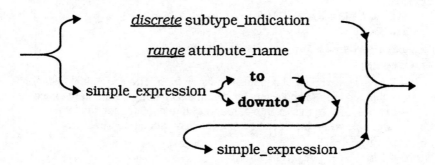

Used In

"Sequential statement" on page 188

Examples

```
LFOR: for J in 0 to MAXADDR – 1 loop
    L_UP_DATA(J) <= (others => '0');
end loop LFOR;
```
    -- This is an example of a for-loop scheme. J takes on values from the
    -- specified discrete range one at a time. The for-loop identifier J has an
    -- implicit declaration and therefore need not be explicitly declared. A for-loop
    -- identifier cannot be assigned a value within the for loop.
    -- LFOR is the loop label.

```
while FOUND = FALSE and ADDR <= MAXADDR loop
    if MOD_NAME = FIFO(ADDR) then
        FOUND := TRUE;
    else
        ADDR := ADDR + 1;
    end if;
end loop;
```
    -- This is an example of while-loop scheme.
    -- Sequential statements within loop execute repetitively as long as the while
    -- condition is true. This loop has no label.

```
SUM := 0; J := 0;
LP: loop
    SUM := SUM + J;
    J := J + 1;
    exit LP when J = 10;
end loop LP;
```
-- This is an example of a loop ... end loop form. The sequential statements
-- within the loop execute repetitively until some other statement causes it to
-- exit the loop. In this example, the exit statement causes the loop to
-- terminate when the specified exit condition becomes true.

```
for K in 1 to TOTAL_VECTORS loop
    READLINE (PAT_IN, IN_BUF);
    WRITELINE (PAT_OUT, IN_BUF);
end loop;
```
-- Copies TOTAL_VECTORS number of lines from file PAT_IN to
-- file PAT_OUT.

```
WFOR: while not ENDFILE(PAT_IN) loop
    READLINE (PAT_IN, IN_BUF);
    WRITELINE (PAT_OUT, IN_BUF);
end loop WFOR;
```
-- Copies all lines from file PAT_IN to file PAT_OUT.

```
PFOR: loop                      -- Assumed that file PAT_IN is non-empty.
    READLINE (PAT_IN, IN_BUF);
    COUNT := COUNT + 1;
    if COUNT mod 2 = 0 then
        WRITELINE (PAT_OUT, IN_BUF);
    end if;
    exit when ENDFILE(PAT_IN);
end loop PFOR;
```
-- Writes out alternate lines from file PAT_IN to file PAT_OUT.

2.56 *Name*

A name is one of the following shown in the syntax.

Syntax

name

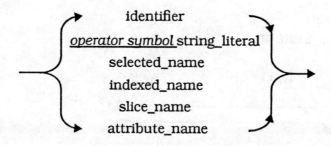

identifier
operator symbol string_literal
selected_name
indexed_name
slice_name
attribute_name

Used In

Almost every construct.

Examples

RESET	-- Identifier: for example, a signal name.
ALU	-- Identifier: for example, an entity name.
"+", "and"	-- Operator string literals.
STD_LOGIC_1164.STD_LOGIC	-- Selected name.
SC2X (FIRSTD, 5)	-- Indexed name.
ADDR (16 **downto** 13)	-- A slice name, specifying a slice of -- the array ADDR.
ARG'RANGE, CLK'EVENT	-- Attribute names.

2.57 *Next statement*

A next statement can only appear inside a loop statement. It causes the current loop iteration of the specified loop to terminate and starts the next iteration, that is, control jumps back to the beginning of the specified loop. If no loop label is specified, by default the innermost loop is assumed.

Syntax

next_statement

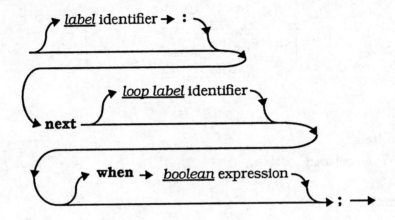

Used In

"Sequential statement" on page 188

Examples

N1: **next**;

 -- When executed will terminate the current iteration of the innermost loop
 -- and the next iteration of the loop will start.
 -- N1 is the label for the next statement.

next LOP;

 -- When executed will terminate the current iteration of the loop with
 -- the specified label, LOP. The loop need not be the innermost loop.

next LAZY **when** A <= 10;

> -- If the boolean condition is false, that is, A is greater than 10, then
> -- no action takes place. If the boolean expression is true, then the
> -- current iteration of the loop labeled LAZY is terminated and the next
> -- iteration of loop LAZY begins.

N2: **next when** LOOP_COUNT **mod** 5 = 0;

> -- Will terminate current iteration of innermost loop if the specified condition is
> -- true. Otherwise execution continues with the statement following this
> -- statement.

LAB_B: **for** K **in** 1 **to** 15 **loop**
    next LAB_B **when** CT_INC(K) <= 0;
    TOTAL_CT := TOTAL_CT + CT_INC(K);
end loop LAB_B;

> -- Loop adds only the positive numbers in array, CT_INC.
> -- When an element of CT_INC has a non-positive value, the next iteration of
> -- loop is performed. For example, if say K is 3 and CT_INC(K) is –15, then
> -- when the **next** statement is executed, the assignment to TOTAL_CT is
> -- skipped and the next iteration of the for-loop with K as 4 is performed.

2.58   *Null statement*

A null statement causes no action to take place.

Syntax

null_statement

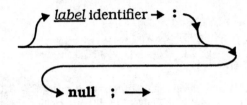

Used In

"Sequential statement" on page 188

Examples

NL1: **null**;
-- NL1 is the statement label.

null;
-- This null statement has no label.

case NEXT_STATE **is**
    when 0 **to** 8 => PROJ_OUT := MATRIX (NEXT_STATE) (2);
    when 9 | 13 => PROJ_OUT := MATRIX (NEXT_STATE) (0);
    when others => **null**;
end case;
-- Use of a null statement in a case statement.

2.59   *Package body*

For every package body, there must be a package declaration with the same package name. If a package declaration has a deferred constant declaration or a subprogram declaration, then a corresponding package body must be present and must contain the complete constant declaration or the subprogram body.

A package body may additionally contain other declarations as well. However, their scope is restricted to the package body, and they cannot be exported to other design units.

Syntax

package_body

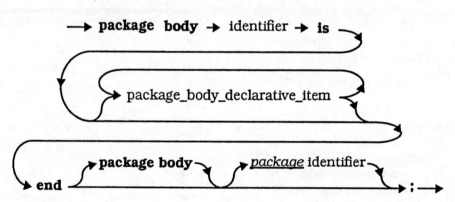

package_body_declarative_item

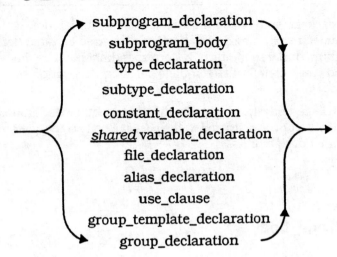

subprogram_declaration
subprogram_body
type_declaration
subtype_declaration
constant_declaration
shared variable_declaration
file_declaration
alias_declaration
use_clause
group_template_declaration
group_declaration

Used In

"Design file" on page 90

Examples

```
package body PACKAGE_NAME is
    -- subprogram declarations
    -- subprogram bodies
    -- type and subtype declarations
    -- constant declarations
    -- shared variable declarations
    -- file and alias declarations
    -- use clauses
    -- group template and group declarations
end package body PACKAGE_NAME;

package body LOGIC4_PKG is
    function RESOLVED (DRIVERS: LOGIC4_VECTOR)
        return LOGIC4 is
        -- Local declarations here.
    begin
        -- Body of function.
    end function RESOLVED;
```

```
        constant STROBE_DELAY: TIME := 15 ns; -- Complete constant declaration.
end package body LOGIC4_PKG;



package body CONV_FUNCTIONS is
    function TO_INTEGER (OPD: BIT_VECTOR) return INTEGER is
    begin
        -- Functionality here.
    end function TO_INTEGER;

    function TO_INTEGER (OPD: UNSIGNED) return INTEGER is
    begin
        -- Functionality here.
    end function TO_INTEGER;

    function TO_INTEGER (OPD: SIGNED) return INTEGER is
    begin
        -- Functionality here.
    end function TO_INTEGER;

    function TO_INTEGER (OPD: STD_LOGIC_VECTOR)
        return INTEGER is
    begin
        -- Functionality here.
    end function TO_INTEGER;
end package body CONV_FUNCTIONS;

package body EXTRA is
    procedure PRESET_CLEAR (signal S: out INTEGER;
                                    PC_VALUE: in INTEGER) is
    begin
        S <= PC_VALUE;
    end procedure PRESET_CLEAR;
end; -- Keywords package body, and the package name are optional after
    -- keyword end.
```

2.60   *Package declaration*

A package is a place to collect commonly used declarations and subprograms. These declarations and subprograms can then be exported to other design units.

A package declaration declares the interface to a package. It may contain deferred constant declarations and subprogram declarations, among other declarations. If the package declaration contains a deferred constant declaration or a subprogram declaration, then a package body is required.

Syntax

package_declaration

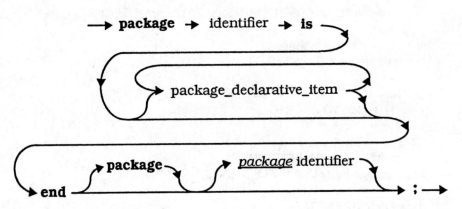

package_declarative_item

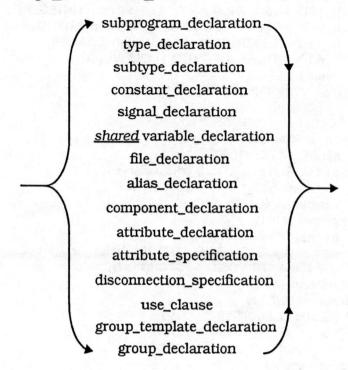

subprogram_declaration
type_declaration
subtype_declaration
constant_declaration
signal_declaration
shared variable_declaration
file_declaration
alias_declaration
component_declaration
attribute_declaration
attribute_specification
disconnection_specification
use_clause
group_template_declaration
group_declaration

Used In

"Design file" on page 90

Examples

```
package LOGIC4_PKG is
    type LOGIC4 is ('0', '1', 'X', 'Z');
    type LOGIC4_VECTOR is array (NATURAL range <>) of LOGIC4;

    function RESOLVED (DRIVERS: LOGIC4_VECTOR)
        return LOGIC4;
    subtype RESOLVED_LOGIC4 is RESOLVED LOGIC4;
    constant STROBE_DELAY: TIME;       -- Deferred constant declaration.
end package LOGIC4_PKG;
    -- A package body is required for this package declaration since it has
    -- a function declaration.
```

```
package CONV_FUNCTIONS is
    function TO_INTEGER (OPD: BIT_VECTOR) return INTEGER;
    function TO_INTEGER (OPD: UNSIGNED) return INTEGER;
    function TO_INTEGER (OPD: SIGNED) return INTEGER;
    function TO_INTEGER (OPD: STD_LOGIC_VECTOR)
        return INTEGER;
end package CONV_FUNCTIONS;

package GENERIC_COMP_DECL is
    component GAND is
        generic (N: POSITIVE);
        port (A, B: in STD_LOGIC_VECTOR(1 to N);
            Z: out STD_LOGIC_VECTOR(1 to N));
    end component GAND;
    component GOR is
        generic (N: POSITIVE);
        port (A, B: in STD_LOGIC_VECTOR(1 to N);
            Z: out STD_LOGIC_VECTOR(1 to N));
    end component GOR;
    component GNOT is
        generic (N: POSITIVE);
        port (A: in STD_LOGIC_VECTOR(1 to N);
            Z: out STD_LOGIC_VECTOR(1 to N));
    end component GNOT;
end package GENERIC_COMP_DECL;

package EXTRA is
    procedure PRESET_CLEAR (signal S: out INTEGER;
                                PC_VALUE: in INTEGER);
end; -- Keyword package and package name are optional after keyword end.
```

2.61 *Physical literal*

A physical literal is a value of a physical type. Such a value represents a measurement of some physical quantity.

If no value is specified with a unit name, the integer literal 1 is assumed.

A physical literal also has a position number associated with it, which is the number of base units.

Syntax

physical_literal

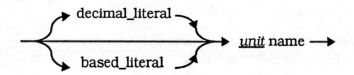

Used In

"Literal" on page 150
"Physical type declaration" on page 167

Examples

20 Volts

    -- A value of a physical type (say VOLTAGE) that has Volts as a unit.

50 Kmeters

    -- A value of a physical type (say DISTANCE) that has Kmeters as a unit.

5 nA

    -- Note that a space must be present between the literal and the unit.

pF

    -- Implies 1 pF.

6.2 ms

>   -- Note that real values can also be used. The position number of this literal is
>   -- the largest base unit that is not greater than the specified value. Assuming
>   -- this literal belonged to a physical type with a base unit of "ms", then the
>   -- position number of this literal is 6.

−4 STEP_1

>   -- Negative values can also be used, as long as it is in the range of the
>   -- physical literal to which this unit belongs.

ns

>   -- Implies 1 ns.

2#1100# nF

>   -- Equivalent to $(1100)_2 = 12$ nF.

6.21 ns

>   -- Equivalent to 6210 ps.

5.6493 ps

>   -- Rounded off to 5649 fs. Value is converted to nearest base unit.

2.62 *Physical type declaration*

A physical type declaration declares a type that contains values that represent measurement of some physical quantity. The type declares the units and the range to be used in specifying such a measurement.

The only predefined physical type is TIME, and the only predefined physical subtype is DELAY_LENGTH.

Syntax

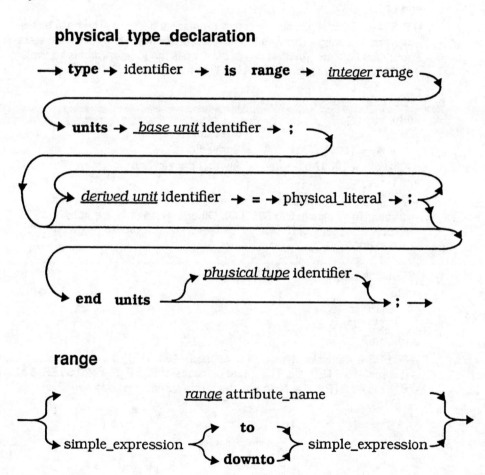

Used In

"Type declaration" on page 219

Examples

type DISTANCE **is range** 0 **to** 2E5
    units
        meters;   -- Base unit.
        Kmeters = 1000 meters;   -- Derived unit.
    end units;
-- DISTANCE is the physical type that represents distance. The base unit is
-- meters. The range specifies the range of the base units which is 0 meters
-- to 200000 meters. Another unit of this type is the Kmeters that is equivalent
-- to 1000 meters. Thus, this physical type has two units defined.

type VOLTAGE **is range** 0 **to** 5E6
    units
        mVolts;   -- Base unit.
        Volts = 1000 mVolts;   -- Derived unit.
        KVolts = 1000 Volts;   -- Another derived unit.
    end units;
-- VOLTAGE is a physical type that represents voltage. The base unit is
-- mVolts. The range is 0 to 5000000. Other derived units are Volts and
-- KVolts. The range of KVolts is thus 0 to 5 and the implied range of Volts
-- is 0 to 5000.

type STEPS **is range** +100 **downto** −15
    units
        STEP_1;
        STEP_5 = 5 STEP_1;
    end units;
-- STEPS is a physical type with a range of +100 STEP_1's
-- downto −15 STEP_1's. The range of values of STEP_5 is +20 STEP_5's
-- downto −3 STEP_5's. Note that the range can also include negative
-- values.

2.63   *Procedure call statement*

A procedure call statement calls a procedure when executed. After the procedure has completed execution, execution continues with the statement following the procedure call.

An actual parameter must be a signal, variable, or a file if the formal parameter is a signal, variable, or a file respectively. The actual can be an expression for a formal of constant class. Values of a variable or a constant are passed in from actuals to formals and vice versa; for signals of mode **in** or **inout**, the entire signal is copied in; for signals of mode **out** or **inout**, the driver for the signal is also copied in; for a file, only a reference to the file is passed in.

Syntax

procedure_call_statement

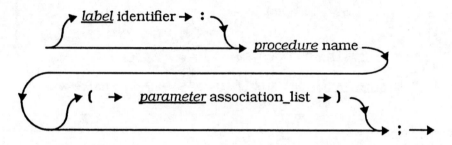

association_list

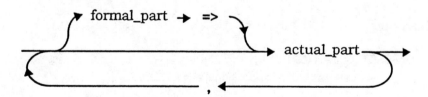

formal_part^{mod}

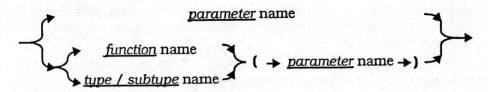

actual_part^{mod}

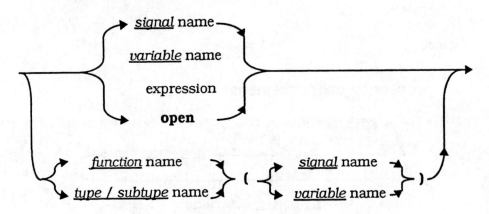

Used In

"Sequential statement" on page 188

Examples

PRESET_CLEAR (DFF, '1');
-- Actuals passed using positional association.

P1: COMPUTE_DELAY (OLD_VALUE, NEW_VALUE, DELAY);
-- P1 is the label for the procedure call statement.

ALU (OPD1 => A, OPD2 => B, OPCODE => ADD, RESULT => SUM);
-- Named association used.

NAND_BEHAVIOR (A => **open**, B => S1, Z => ACK);
>
> -- Since input parameter A has no actual value, the parameter declaration for
> -- A in the procedure declaration or body must have an explicit default value
> -- specified. For example,

procedure NAND_BEHAVIOR (A, B: **in** BIT := '0'; Z: **out** BIT);

> -- Here is another procedure call:

NAND_BEHAVIOR (A => STA, Z => PROP);
>
> -- In this case, since no actual value is specified for the formal
> -- parameter B, the default value for the parameter is used, which is '0'.

AND_COMPUTE (A0 => D(0) **xor** D(1), A1 => D(2) **xor** D(3), Z => CTRH);
>
> -- Actuals can also be expressions.

ADD_WITH_CARRY (A0 => TO_STDLOGIC(BUN),
                A1 => TO_STDLOGIC(FUN),
                CIN => TO_STDLOGIC(SUN),
                TO_BIT(SUM) => DON, TO_BIT(CARRY) => ROD);
>
> -- TO_STDLOGIC and TO_BIT are conversion functions. A0, A1, and
> -- CIN are inputs while SUM and CARRY are outputs. For input parameters,
> -- the return value of the function TO_STDLOGIC is passed into the
> -- procedure. For output parameters, the return value of the function TO_BIT
> -- is passed to the corresponding actuals.

2.64   *Process statement*

A process statement models the sequential behavior of a part of a design. The sequential behavior is expressed using sequential statements.

A process is always in a suspended state or in an execution state. Every process is executed once during the initialization phase of simulation. A process suspends either due to the presence of a sensitivity list or because of a wait statement in the process. If a process has a sensitivity list, then the process executes every time there is an event on any signal in the sensitivity list and suspends after executing the last statement in the process.

If a process statement has a sensitivity list, then no wait statements are allowed in the process. If a process does not have a sensitivity list, then the process must have one or more wait statements.

If a process is a postponed process, defined by the presence of the keyword **postponed**, then the process will never trigger at delta times, for example at time 15 ns+1Δ, but will trigger only at the end of a time step. Also, when a postponed process executes, it is an error if it causes new events to occur after a delta time.

Declarations appearing in a process statement are local only to the process.

Syntax

process_statement

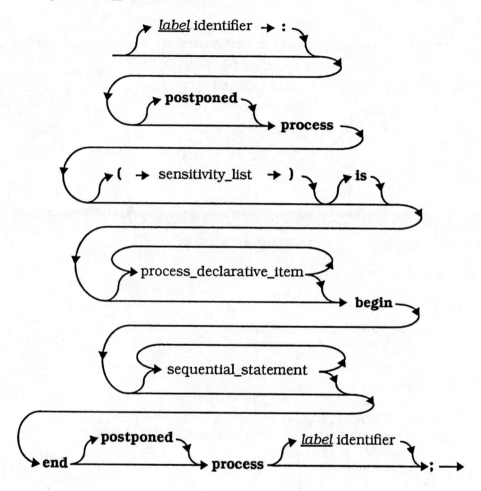

sensitivity_list

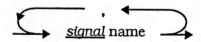

process_declarative_item

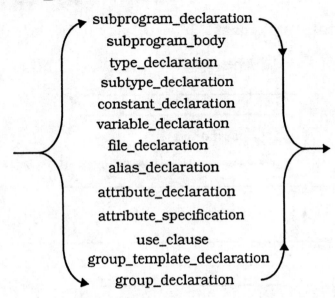

subprogram_declaration
subprogram_body
type_declaration
subtype_declaration
constant_declaration
variable_declaration
file_declaration
alias_declaration
attribute_declaration
attribute_specification
use_clause
group_template_declaration
group_declaration

Used In

"Concurrent statement" on page 66
"Entity declaration" on page 95

Examples

PROCESS_LABEL: **process** (SENSE1, SENSE2, SENSE3, SENSE4) **is**
    -- Process will trigger on events on signals SENSE1, SENSE2,
    -- SENSE3 or SENSE4.
    -- Local declarations here.
begin
    -- Any number of sequential statements here, but no wait statements
    -- allowed since this process has a sensitivity list.
end process PROCESS_LABEL;

```
process is
     -- No sensitivity list for this process.
     -- Local declarations here.
begin
     -- Any number of sequential statements, including any number
     -- of wait statements. At least one wait statement must be present.
end process;     -- Process label is optional.
```

```
P1: process (A, B)                    -- Keyword is is optional.
     -- P1 is the process label.
     -- A and B are the signals in the sensitivity list.
begin
     -- There is only one signal assignment statement in this process:
     Z <= A or B after 10 ns;
end process P1;
```
     -- A process statement with a sensitivity list has an equivalent process with
     -- a wait statement. The equivalent process for the above example is:

```
P1: process                -- No sensitivity list.
begin
     Z <= A or B after 10 ns;
     wait on A, B;          -- Wait statement derived from sensitivity list in
                            -- original process.
end process;
```

```
PQ: postponed process (DRY) is
begin
     if DRY = '0' then
          CLK <= '1' after 5 ns;
     else
          CLK <= '0' after 10 ns;
     end if;
end process PQ;
```
     -- A postponed process does not trigger at any delta time. These delta time
     -- events are postponed to the end of the time step. In this example, if DRY
     -- had events at times 14 ns, 14 ns + 1Δ, 14 ns + 2Δ, 18 ns, 18 ns + 1Δ,
     -- 21 ns + 2Δ, then the process will only trigger after the end of all deltas, that
     -- is, after all deltas at times 14 ns, 18 ns, and 21 ns.

```
CLOCK: process is
    constant OFF_DELAY: TIME := 5 ns;
    constant ON_DELAY: TIME := 10 ns;
    constant MAX_CLOCK_TIME: TIME := 900 ns;
begin
    RCLK <= '0';
    wait for OFF_PERIOD;
    RCLK <= '1';
    wait for ON_PERIOD;
    if NOW < MAX_CLOCK_TIME then
        report "Simulation completed successfully.";
        wait;            -- Suspend process indefinitely.
    end if;
end process;
    -- This process statement will suspend at each of the wait statements for the
    -- specified times.
```

2.65   *Qualified expression*

A qualified expression is used to explicitly state the type or subtype of an expression. It does not imply type conversion.

It is most commonly used to qualify the type of a literal or an aggregate.

Syntax

qualified_expression

Used In

"Allocator" on page 22
"Simple expression" on page 195

Examples

BIT'('0')
    -- The literal '0' belonging to the BIT type. If the type were not specified, then
    -- the surrounding context is used to determine the type of the literal. This
    -- may be confusing in some cases, and it may be useful to explicitly state
    -- the type that the value belongs to.

CHARACTER'('1')
    -- The literal '1' belonging to the CHARACTER type.

STD_LOGIC_VECTOR'("1100")
    -- A vector of type STD_LOGIC_VECTOR.

MVL_VECTOR'(0 => 'U', 1 **to** 4 => '0', 5 | 7 => 'U', 6 => '1')
    -- An aggregate of type MVL_VECTOR.

BIT_VECTOR'(A+B)
    -- The expression result of type BIT_VECTOR.

2.66   *Record type declaration*

A record is a collection of one or more elements, these may possibly be of different types. The format of a record is declared using a record type declaration.

Syntax

record_type_declaration

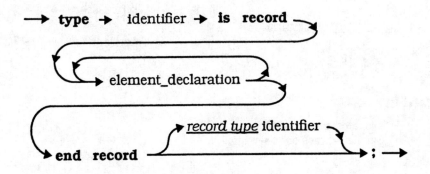

element_declaration

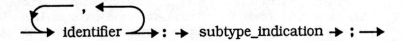

Used In

"Type declaration" on page 219

Examples

```
type MICRO_BUS is
    record
        ADDRESS     : STD_ULOGIC_VECTOR (ADDR_WIDTH-1 downto 0);
        DATA        : STD_ULOGIC_VECTOR (DATA_WIDTH-1 downto 0);


        READ,
        WRITE       : STD_ULOGIC;
    end record;
    -- MICRO_BUS is a record type containing four elements.

type DIGIT_TYPE is ('0', '1', '2', '3', '4', '5', '6', '7', '8', '9');
    -- DIGIT_TYPE is an enumeration type.
type SS_TYPE is array (1 to 9) of DIGIT_TYPE;
    -- SS_TYPE is a constrained array type.
type EMPLOYEE is   -- EMPLOYEE is the record type.
    record
        NAME            : STRING (0 to 15);
        LAST_SALARY,
        CURR_SALARY : INTEGER range 0 to 100_000;
        SS_NO           : SS_TYPE;
        YRS_OF_EXP    : INTEGER range 0 to 100;
    end record EMPLOYEE;
    -- This record type has five elements of various types.

type TRIPLETS is
    record
        A, B, C: BIT_VECTOR (0 to 3);
    end record;

type COMPLEX is
    record
        REAL_NUM, IMAG_NUM : REAL;
end record COMPLEX;
    -- This record type has two elements, both of type REAL.
```

2.67   *Report statement*

A report statement is similar to an assertion statement with an assertion value of false. That is, a report statement when executed causes the report message to be printed and passes the specified severity level to the simulator. Based on the severity level, a simulator may abort further processing, or resume simulation or print out certain diagnostic information before continuing simulation.

Syntax

report_statement

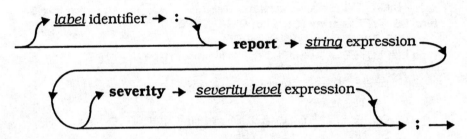

Used In

"Sequential statement" on page 188

Examples

    R1: **report** "This code should not have been entered!"
        severity NOTE;
        -- R1 is the label of this report statement. This statement when executed
        -- causes the report message to be printed and passes the severity level of
        -- NOTE to the simulator.

    R2: **report** "A new event occurred on clock.";
        -- The default severity level of NOTE is used.

    report "This IF condition can never be true. Aborting ..."
        severity FAILURE;
        -- This report statement has no label.

2.68   *Return statement*

A return statement causes the enclosing procedure or function to terminate. If a return statement appears inside a procedure, then no expression is allowed in the return statement. Every function must have at least one return statement in which the return statement must have an expression and the type of the expression must match the return type of the function.

Syntax

return_statement

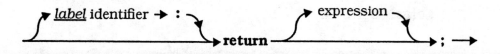

Used In

"Sequential statement" on page 188

Examples

return;

    -- Allowed only inside a procedure. Causes the procedure to terminate.

R1: **return** A + B;

    -- The value of A+B is returned from the function.
    -- R1 is the label for the return statement.

return RANDOM;

    -- The value of the variable RANDOM is returned from the function.

2.69   *Selected name*

A selected name represents an item that is declared within another item or within a design library, for example, elements of a record, items declared in a package, items present in a library, items declared inside a process, block or loop. A selected name consists of a prefix, followed by a period, followed by the explicit item being referenced. The prefix may be a loop label, process label, block label, package name, library name, record name, or a function that returns a record.

Syntax

selected_name

prefix

Used In

"Name" on page 155
"Use clause" on page 223

Examples

STD.STANDARD.BIT

>   -- Type BIT defined in package STANDARD that resides in design
>   -- library STD.

SIG_BUS.READ

>   -- Element READ of record SIG_BUS.

EMPLOYEE.PAY_ROLL

>   -- Element PAY_ROLL of record EMPLOYEE.

CMOS.**all**

>   -- All primary units residing in CMOS design library.

S1.SAR

>   -- Signal SAR declared in block S1.

BIT_ARITH.PRESET_CLEAR

>   -- Procedure PRESET_CLEAR declared in package BIT_ARITH.

MISC.LOGIC_ARITH."+"

>   -- Overloaded function operator "+" declared in package LOGIC_ARITH
>   -- residing in library MISC.

STANDARD.'A'

>   -- The character 'A' declared in package STANDARD.

"+" (A, B). REAL_PART

>   -- "+" is a function that operates on records that hold complex values and
>   -- returns such a record. REAL_PART is an element of the record.

DFG_PTR.**all**

>   -- Assuming that DFG_PTR is an object of an access type, then this name
>   -- represents the object pointed to by DFG_PTR.

2.70 *Selected signal assignment statement*

A selected signal assignment statement behaves very much like a case statement. However, a selected signal assignment statement is a concurrent statement while a case statement is a sequential statement. The selected signal assignment statement executes whenever an event occurs on a signal used in the select expression or in any of the waveforms. Based on the value of the select expression, the appropriate waveform is evaluated and assigned to the target. All values in the type of the select expression must be covered, and each must be covered exactly once.

Syntax

selected_signal_assignment_statement

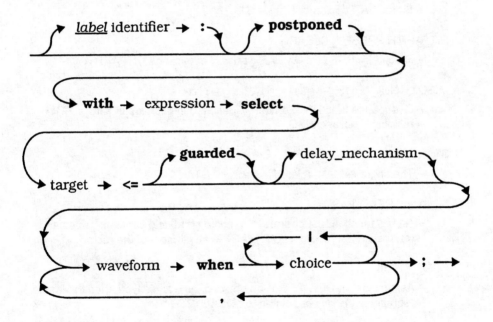

delay_mechanism

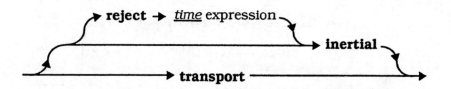

target

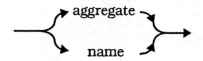

waveformmod

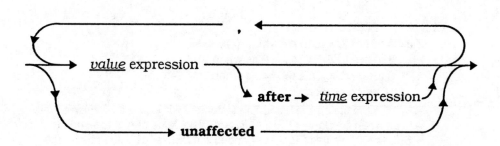

choice

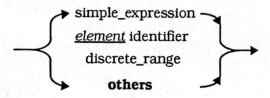

discrete_range

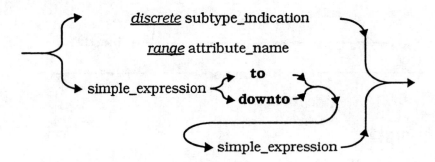

Used In

"Concurrent statement" on page 66

Examples

MUX: **with** SEL(0 **to** 1) **select**
    Z <= A **after** MUX_DELAY **when** "00",
        B **after** MUX_DELAY + 2 ns **when** "01",
        C **after** MUX_DELAY + 4 ns **when** "11",
        'U' **when others**;

    -- MUX is the statement label. The statement will execute whenever there
    -- is an event on signals SEL, A, B, or C. The value of A is assigned to Z if
    -- SEL is "00", value of B is assigned to Z if SEL is "01", and so on. The
    -- keyword **others** covers all remaining values of SEL(0 **to** 1).

PATTERNS: **with** TEST **select**
    PAT_OUT <= '0', '1' **after** 2 ns, '0' **after** 5 ns **when** 1,
                '**U**', '1' **after** 10 ns, '0' **after** 15 ns **when** 2,
                'U', '1' **after** 10 ns, 'U' **after** 12 ns, '0' **after** 20 ns
                    when 3;

    -- Assuming that TEST has values only in the range 1 to 3. This statement will
    -- execute only when an event occurs on signal TEST.

```
FSM: with CURRENT_STATE select
     NEXT_STATE <= S1 when S0,
                   S2 when S1,
                   S3 when S2,
                   S0 when others;
```

For examples of **postponed, guarded, unaffected, reject, inertial** and **transport,** see "Conditional signal assignment statement".

2.71   *Sequential statement*

A sequential statement executes sequentially with respect to other surrounding statements. It can appear within a process statement, or in a function or procedure body.

Syntax

sequential_statement

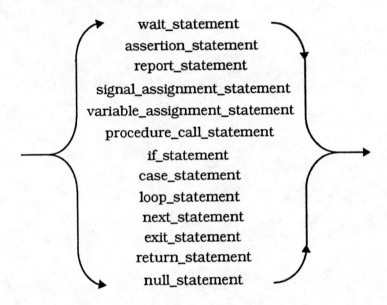

wait_statement
assertion_statement
report_statement
signal_assignment_statement
variable_assignment_statement
procedure_call_statement
if_statement
case_statement
loop_statement
next_statement
exit_statement
return_statement
null_statement

Used In

Examples

See respective statements.

2.72   *Signal assignment statement*

A signal assignment statement is used to assign a value to a signal or to an aggregate set of signals. When a signal assignment statement is executed, the waveform on the right-hand-side of the assignment is evaluated, and this waveform is assigned to the target signal or signals.

The waveform may be composed of multiple waveform elements, where each waveform element specifies the value and the time after which the value is to appear. If no time expression is present, a delta delay is assumed.

The keyword **transport** indicates that the transport delay mode should be used. The default is the inertial delay mode. In the transport delay mode, the waveform formed from the right-hand-side expression appears on the target after the specified delay. In the inertial delay mode, the value of the waveform on the right-hand-side must be stable for at least the specified pulse rejection limit (default rejection limit is the inertial delay of the first waveform element) before it is allowed to propagate to the left-hand-side after the specified delay. The pulse rejection time must not be greater than the delay of the first waveform element.

The value **null**, when assigned to a guarded signal, is used to disconnect the driver of the signal.

Syntax

signal_assignment_statement

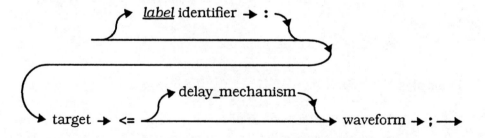

target

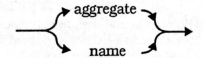

delay_mechanism

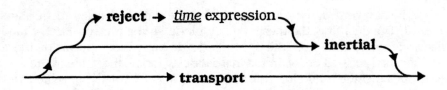

waveform^{mod}

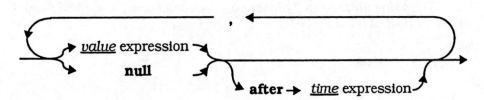

Used In

"Sequential statement" on page 188

Examples

S1: RESET <= '0', '1' **after** 15 ns, '0' **after** 100 ns;

    -- This signal assignment has 3 waveform elements. Delay is inertial delay.
    -- The first '0' is assigned after delta delay. S1 is the label for the signal
    -- assignment.

Z <= **inertial** A **nand** B **after** 5 ns, A **xor** B **after** 10 ns;

    -- Inertial delay is explicitly specified. Pulse rejection limit is same as inertial
    -- delay of first waveform element which is 5 ns.

L2: DELAYED_S1 <= **transport** S1 **after** TRANSPORT_DELAY;

    -- Transport delay is used. L2 is the statement label.

(A, B, C) <= BIT_VECTOR'("001");

    -- A and B get '0' after delta delay, C gets '1' after delta delay. The left-hand-
    -- side is an aggregate.

(P_Q, P_QN, K) <= P_TABLE(NEXT_STATE) **after** 5 ns;

    -- Assume that P_TABLE is a 3-bit constant. Leftmost bit is assigned to signal
    -- P_Q, next bit assigned to P_QN, and the last bit is assigned to K. The left-
    -- hand-side is again an aggregate.

Z <= **reject** SPIKE_WINDOW **inertial** A **xor** B **after** GET_RISE_FALL (A, B);

    -- If value of (A **xor** B) changes in the SPIKE_WINDOW period, then change
    -- is not propagated to Z. If the value of (A **xor** B) is stable for the
    -- SPIKE_WINDOW period, then the value appears on Z after the delay
    -- returned by the function GET_RISE_FALL.

SDTY <= **null**;

    -- When executed, the driver for signal SDTY of this process is disconnected
    -- after a delta delay. SDTY must be a guarded signal.

L3: SAM <= '0', '1' **after** 2 ns, **null after** 5 ns, '0' **after** 11 ns;

    -- If this statement is executed at time 20 ns, then the driver to SAM in the
    -- block in which this statement appears is disconnected after (20+5) ns.
    -- SAM must be a guarded signal.

DST <= **reject** 10 ns **inertial** RDY **after** 5 ns;

    -- This is an error since pulse rejection limit must not be greater than the
    -- delay of the first waveform element.

WADDROUT <= TO_X01ZVECTOR (BIN_WADDR, WADDROUT'LENGTH);

    -- Uses delta delay. Next two examples also use delta delay.

READFIFO <= '1';

CNT <= CNT + 1;

2.73 Signal declaration

A signal declaration declares one or more signals of a specified type. A signal has the property that it can be assigned a value using a signal assignment statement and that it can only be assigned a future value, i.e., the current value of the signal can never be changed. Also a signal retains the current value, the future set of values that are scheduled to appear on the signal, and its past value.

If the signal's type declaration contains the name of a resolution function, then the signal is said to be a resolved signal.

The signal declaration may optionally specify a signal kind, namely, **register** or **bus**. Such a signal is called a *guarded signal*, and it must also be a resolved signal. The **register** kind of signal differs from that of the **bus** kind in that if all drivers to a **register** signal are turned off, the signal will retain the last value that appears on that signal; whereas in a **bus** kind of signal, if all drivers are turned off, the resolution function is called to determine the effective value of the signal. A guarded signal may be assigned values under the control of a guard expression in a guarded signal assignment. When the guard expression is false, the driver for the guarded signal gets disconnected. The disconnect time is specified using a disconnection specification.

The signal declaration may also optionally specify an initial value. This is the initial value that is assigned to the signal during the initialization phase of simulation. If an initial value is not specified explicitly, then the initial value is the leftmost value of the signal type, that is, *signal_type*'LEFT. If the signal type is a composite type (that is, an array or a record type), then each element of the composite type gets the leftmost value of the corresponding element type.

Syntax

signal_declaration

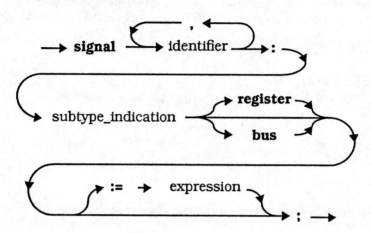

Used In

"Architecture body" on page 24
"Block statement" on page 42
"Entity declaration" on page 95
"Generate statement" on page 120
"Package declaration" on page 162

Examples

signal RDY, CLR: BOOLEAN;
   -- RDY and CLR are two signals of type BOOLEAN and their initial values are
   -- both FALSE (since FALSE is the leftmost value of type BOOLEAN).

signal CARRY_BUS: STD_LOGIC_VECTOR (0 **to** 3) := "0000";
   -- Signal CARRY_BUS has the initial value of "0000" when simulation starts.

signal ACK: WIRED_AND MVL **register**;
   -- ACK is a **register** kind of guarded signal of type MVL. WIRED_AND is the
   -- name of the resolution function used to resolve values from
   -- multiple drivers of signal ACK.

signal BUS_HOLD1, BUS_HOLD2:

A_TO_D.DIGITIZE INTEGER **range** −5 **to** +5;

-- BUS_HOLD1 and BUS_HOLD2 are resolved signals of type INTEGER
-- with a range constraint of −5 to +5. The resolution function associated with
-- the signals is the function DIGITIZE that resides in the package A_TO_D.

signal WADDR: INTEGER **range** 5 **downto** 1;

signal FREQ: FIFOTYPE;

-- FIFOTYPE is a user-defined type that has been defined elsewhere.

2.74   *Simple expression*

A simple expression is an expression involving mainly the arithmetic operators. This also includes the & (concatenation) operator and the logical **not** operator. The predefined operators that can be used in a simple expression are shown in the following table.

Operator	Left operand type	Right operand type	Result type
Addition operators: +, −	Numeric type	Same type	Same type
&	Array or element type	Array or element type	Array type
Sign operators: +, −		Numeric type	Same type
Multiplying operators: *, /	Numeric type	Same type	Same type
mod, rem	Integer type	Same type	Same type
Miscellaneous operators: **	Numeric type	INTEGER	Same as left
abs		Numeric type	Same type
not		BIT/BOOL-EAN or 1D array of BIT/BOOL-EAN	Same as left

The operators in the above table are categorized in increasing precedence. Operators in each category have the same precedence. The miscellaneous operators have the highest precedence. The relational, logical, and shift operators used in the syntax for "expression" have lower precedence than all operators listed in the previous table. A *numeric type* is an integer type, a floating point type, or a physical type.

Syntax

simple_expression

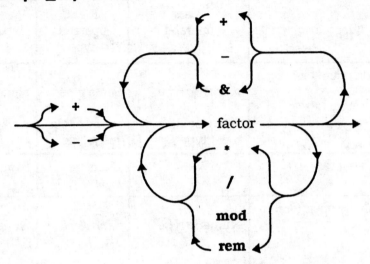

factor

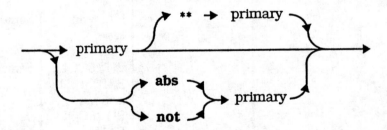

primary

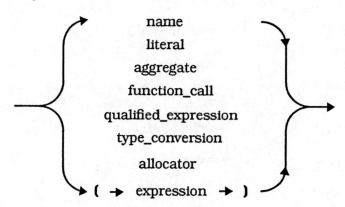

Used In

Examples

PI * (R ** 2) / 2

2 * (LENGTH + WIDTH)

A ** 2 + B ** 2

(COUNTER + 1) **mod** MAXCOUNT

(**not** A)

7 **mod** 4   -- has value 3.

7 **rem** 4   -- also has value 3.

(–7) **mod** 4   -- has value 1

(–7) **rem** 4   -- has value –3

7 **mod** (–4)   -- has value –1

7 **rem** (–4)   -- has value 3

(–7) **mod** (–4)   -- has value –3

(–7) **rem** (–4)   -- has value –3

2.75 *Slice name*

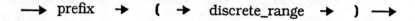

A slice name represents a subrange of an array. It consists of an array name or
a function that returns an array, followed by a range specification.

Syntax

slice_name

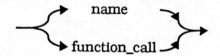

$\longrightarrow$ prefix $\rightarrow$ ($\rightarrow$ discrete_range $\rightarrow$) $\longrightarrow$

prefix

name

function_call

discrete_range

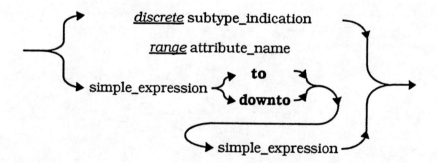

discrete subtype_indication

range attribute_name

simple_expression **to** **downto**

simple_expression

Used In

"Name" on page 155

Examples

ADDRESS (0 **to** 4)

    -- Elements 0 to 4 of the ADDRESS array.

DATA (LO'RANGE)

    -- If LO is an array of range 0 to 3, then elements of DATA at index 0 to 3 are
    -- being referenced.

TO_BITVECTOR(15, 6) (3 **downto** 0)

    -- The 3 through 0 elements of the result of the TO_BITVECTOR function.

PC_VALUE (5 **downto** 10)

    -- Represents a null slice since no elements are represented by this slice.
    -- Such a range is called a null range.
    -- Another example of a null range is "0 **to** −10".

2.76   *String literal*

A string literal represents a value for an array of zero or more characters.

Syntax

string_literal

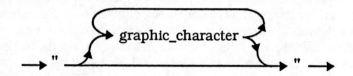

Used In

"Alias declaration" on page 19
"Attribute specification" on page 35
"Literal" on page 150
"Name" on page 155
"Selected name" on page 182
"Subprogram body" on page 202
"Subprogram declaration" on page 208

Examples

"Exiting simulation ..."

"CLK and RESET are both HIGH!"

" "" "

    -- The " character itself.

" "

    -- Null string literal.

"000111XX"

    -- An example of a string of type STD_ULOGIC_VECTOR.

2.77   Subprogram body

A subprogram body defines the computation or algorithm performed when a subprogram is called. It also specifies the interface to the subprogram, that is, the list of parameters that are passed in and out of the subprogram. A subprogram is a procedure or a function.

A function always returns a value, while a procedure can return zero or more values. A function name may also be a operator symbol, for example, when an overloaded operator function is being defined. Function parameters are by default constants, and the only mode allowed is in. Procedure parameters are by default constants of mode in, while out and inout parameters are by default variables. A parameter may optionally have a default expression. The value of the default expression is used if no actual value is specified for the formal parameter or if the keyword open is specified in the subprogram call.

Items declared in a subprogram are local only to the subprogram. Objects declared within a subprogram (variables, constants and files) get created and initialized every time the subprogram is called.

A function can be a pure or an impure function. It is said to be a pure function if it returns the same value each time it is called with the same set of actual values. Otherwise, it is an impure function. For example, the predefined function NOW is an impure function since it returns different values on different invocations. If the keyword impure is present in a function specification, then the function is declared to be an impure function, otherwise it is a pure function.

A subprogram is said to be a foreign subprogram if an attribute specification for the 'FOREIGN attribute appears in the declarative part of the subprogram body. In such a case, the value of the attribute, which is a string, may specify implementation-dependent information. Such an implementation may be a non-VHDL implementation as well. The parameter passing mechanism is not defined for these foreign subprograms.

Syntax

subprogram_body

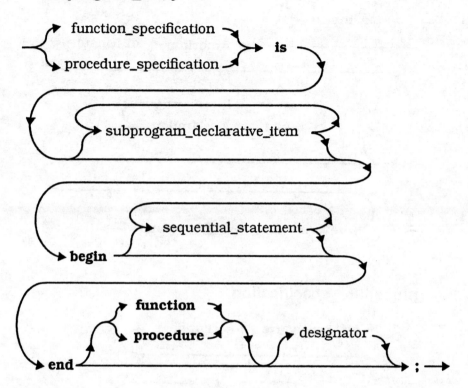

function_specification

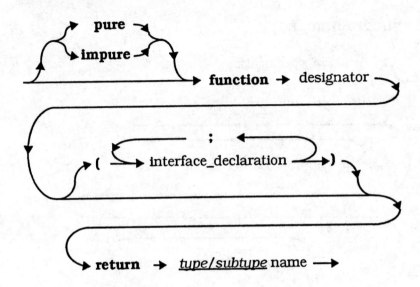

procedure_specification

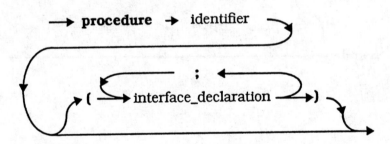

designator

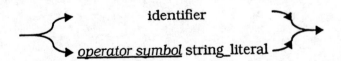

interface_declaration

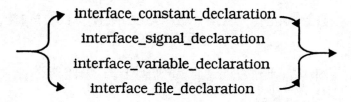

interface_constant_declaration
interface_signal_declaration
interface_variable_declaration
interface_file_declaration

subprogram_declarative_item

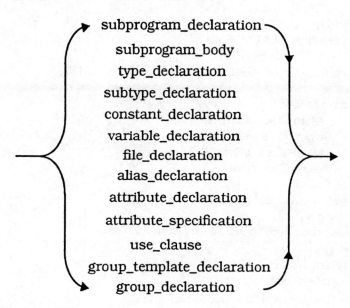

subprogram_declaration
subprogram_body
type_declaration
subtype_declaration
constant_declaration
variable_declaration
file_declaration
alias_declaration
attribute_declaration
attribute_specification
use_clause
group_template_declaration
group_declaration

Used In

"Architecture body" on page 24
"Block statement" on page 42
"Entity declaration" on page 95
"Generate statement" on page 120
"Package body" on page 159
"Process statement" on page 172
"Subprogram body" on page 202

Examples

```
procedure COMPUTE_DELAY (OLD_VALUE, NEW_VALUE: in BIT;
                              DELAY: out TIME) is
begin
    -- Sequential statements appear within a subprogram body.
    if OLD_VALUE = '1' and NEW_VALUE = '0' then
        DELAY := 5 ns;
    elsif OLD_VALUE = '0' and NEW_VALUE = '1' then
        DELAY := 7 ns;
    else
        DELAY := 0 ns;
    end if;
end procedure COMPUTE_DELAY;

function TO_CHAR (OPD: MVL) return CHARACTER is
begin
    case OPD is
        when '0' => return '0';
        when '1' => return '1';
        when 'U' => return 'U';
        when 'Z' => return 'Z';
    end case;
end function TO_CHAR;

pure function "and" (L, R: X01Z) return X01Z is
begin
    return TABLE_AND (L, R);
end function "and";
    -- Example of an overloaded and operator.

procedure GET_PWD (PWD: out STRING) is
    attribute FOREIGN of GET_PWD: procedure is " `pwd` ";
    -- The value of the attribute is interpreted by the host environment when
    -- this procedure is called.
begin
    -- This is a foreign procedure. There need not be any statements in a foreign
    -- subprogram. If present, they are treated in an implementation-dependent
    -- way.
end procedure GET_PWD;
```

```vhdl
procedure AND_WITH_CARRY (A, B, CIN: in STD_LOGIC := '0';
                          signal SUM, COUT: out STD_LOGIC) is
begin
    SUM <= A xor B xor CIN;
    COUT <= A or B or CIN;
end;      -- Keyword procedure and procedure name optional after keyword end.
    -- In this example, the output parameters are explicitly declared to be signals.
    -- Since a default expression is specified for the input parameters, if no actual
    -- value is specified for a parameter in a procedure call, the default value
    -- of '0' is used.
```

2.78   *Subprogram declaration*

A subprogram declaration defines the interface to a subprogram. A subprogram is a function or a procedure. The behavior of a subprogram is defined using a subprogram body.

A function declaration declares the list of input parameters and their types. It also specifies the return type. Only input parameters are allowed in a function, and these are by default constants. A parameter could be explicitly specified to be a signal or a file.

A function may be a pure or an impure function. It is said to be a pure function if the function returns the same value each time it is called with the same set of actual values. Otherwise, it is said to be an impure function. An impure function must include the keyword **impure** in its specification.

A procedure declaration declares the list of parameters, their mode, and their types. Input parameters are by default constants, while output and inout parameters are by default variables. A parameter may also be explicitly specified to be a signal or a file.

An interface parameter may optionally have a default expression. The value of this default expression is used if the actual value of a parameter is not specified or specified as **open** in the subprogram call.

Syntax

subprogram_declaration

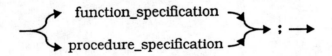

function_specification

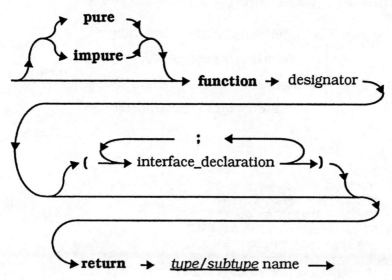

procedure_specification

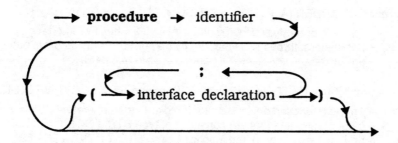

designator

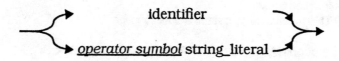

interface_declaration

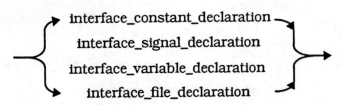

$$
\left\{
\begin{array}{l}
\text{interface\_constant\_declaration} \\
\text{interface\_signal\_declaration} \\
\text{interface\_variable\_declaration} \\
\text{interface\_file\_declaration}
\end{array}
\right\}
$$

Used In

"Architecture body" on page 24
"Block statement" on page 42
"Entity declaration" on page 95
"Generate statement" on page 120
"Package body" on page 159
"Package declaration" on page 162
"Process statement" on page 172
"Subprogram body" on page 202

Examples

procedure ADDER (A, B, CIN: **in** BIT; SUM, COUT: **out** BIT);
    -- This is a procedure declaration for ADDER. It has five interface
    -- parameters: three inputs that are by default constants, and two outputs that
    -- are by default variables. All parameters are of type BIT.

function "-" (L, R: STD_LOGIC_VECTOR) **return** STD_LOGIC_VECTOR;
    -- This is a function declaration for the overloaded operator "-". It has two
    -- input parameters that are constants of type STD_LOGIC_VECTOR,
    -- and the return type of the function is also STD_LOGIC_VECTOR.
    -- Since the input parameter type is an unconstrained array type, the range
    -- for an input parameter is determined from the range of the corresponding
    -- actual specified in a function call. Since the return type is also an
    -- unconstrained array type, the range of the return value is determined by the
    -- range of the expression in a return statement appearing inside the
    -- function body.
    -- This function is by default a pure function.

pure function NAND_REDUCE (A: BIT_VECTOR) **return** BIT;
    -- This is a pure function declaration. It will return the same value each time it
    -- is called with the same value of A.

procedure PRESET_CLEAR (**signal** FF: **inout** STD_LOGIC;
                                   PC_VALUE: STD_LOGIC);

    -- The inout parameter FF is explicitly declared to be a signal. PC_VALUE is
    -- by default a constant of mode **in**.

impure function RANDOM **return** REAL;

    -- This is an impure function declaration. This is because different calls to
    -- RANDOM return different values. Also note that this function has no input
    -- parameters.

procedure DUMP_VEC (**file** VEC_FILE: STD_LOGIC_FTYPE;
                           OUT_VEC: **out** STD_LOGIC_VECTOR);

    -- VEC_FILE is a file with no mode specified. A file parameter must not
    -- have any mode explicitly specified, since files have no mode.

procedure ADD_WITH_CARRY (A, B, CIN: **in** STD_LOGIC := '0';
                                   signal SUM, COUT: **out** STD_LOGIC);

    -- A default value is specified for the input parameters.

2.79 *Subtype declaration*

A subtype declaration defines a subtype. A subtype is a type with a constraint; the type is called the base type. The constraint can possibly be null, in which case the subtype is just an alternate name for the type.

A subtype does not declare a new type. Thus operations defined on the base type are also allowed on the subtype.

Syntax

subtype_declaration

⟶ **subtype** → identifier → **is** → subtype_indication → **;** ⟶

Used In

"Architecture body" on page 24
"Block statement" on page 42
"Entity declaration" on page 95
"Generate statement" on page 120
"Package body" on page 159
"Package declaration" on page 162
"Process statement" on page 172
"Subprogram body" on page 202

Examples

subtype INDEX **is** INTEGER **range** 16 **to** 31;
    -- INDEX is a subtype of type INTEGER with a constraint of integer values
    -- from 16 to 31.

function WIRED_AND (DR: BIT_VECTOR) **return** BIT;
subtype RESOLVED_BIT **is** WIRED_AND BIT;
    -- RESOLVED_BIT is a subtype of type BIT and it also has the resolution
    -- function WIRED_AND associated with it.

type WEEK_DAY **is** (MON, TUE, WED, THU, FRI, SAT, SUN);
subtype WORK_DAY **is** WEEK_DAY **range** MON **to** FRI;
    -- WORK_DAY is a subtype of the enumerated type WEEK_DAY with a range
    -- constraint.

```
type GANG_FF is array (NATURAL range <>)
    of STD_LOGIC_VECTOR (7 downto 0);
subtype REG16 is GANG_FF (0 to 15);
subtype REG8 is GANG_FF (1 to 8);
subtype REG4 is GANG_FF (3 downto 0);
```

    -- REG16 is a subtype of the type GANG_FF with a constraint of indices to
    -- range from 0 to 15. REG4, REG8, and REG16 are subtypes of the same
    -- base type GANG_FF.

```
subtype SYN_INTEGER is INTEGER;
```

    -- SYN_INTEGER is a subtype of INTEGER with no constraint.
    -- Therefore, it is an alternate name for INTEGER.

2.80   *Subtype indication*

The subtype indication is used to specify a type, its constraints, if any, and any resolution function associated with the type.

Syntax

subtype_indication

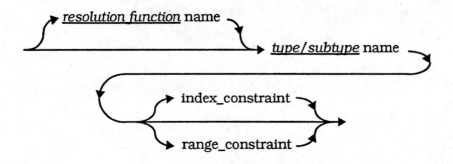

index_constraint

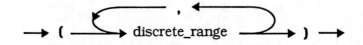

discrete_range

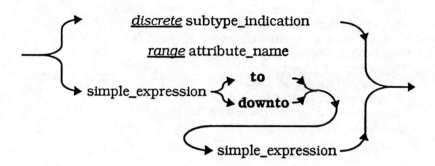

range_constraint

→ **range** → range →

range

Used In

Examples

```
-- The text on the right-hand-side of the ":" symbol are examples
-- of subtype indication:

variable A: INTEGER range 0 to 7;
    -- Range constraint.

constant STROBE_DELAY: TIME;
    -- Physical type.

signal CACHE: BIT_MATRIX(0 to 1023, 7 downto 0);
    -- Index constraint.

type LOGIC4 is ('U', '0', '1', 'Z');
signal DISPENSE: PULL_DOWN LOGIC4;
    -- PULL_DOWN is the name of the resolution function that is associated with
    -- the LOGIC4 type. Therefore, whenever the value of the signal DISPENSE
    -- needs to be resolved, the function PULL_DOWN is called automatically to
    -- determine the effective value of the signal.
```

2.81 *Type conversion*

Type conversion converts a value of a type to that of another type. This is similar to type casting in C programming language. However, VHDL supports very restricted amount of type conversion only. Type conversion is allowed between closely related types. These include between integer types, between real types, and between an integer and a real type. For example, a real can be converted to an integer and vice versa, an integer of one integer type can be converted to an integer of another integer type.

Type conversions are also allowed between arrays that are of closely related types, that is, types in which the element types are the same and the index types are the same or are closely related.

Other forms of type conversion can be achieved by writing suitable functions.

Syntax

type_conversion

$\longrightarrow$ *type / subtype* name $\rightarrow$ **(** $\rightarrow$ expression $\rightarrow$ **)** $\longrightarrow$

Used In

"Simple expression" on page 195

Examples

```
subtype FOUR_BIT_REAL is REAL range 0.0 to 15.0;
variable A: INTEGER;
signal B: FOUR_BIT_REAL;
A := INTEGER (2.6);
    -- Variable A will get the integer value 2.

B <= FOUR_BIT_REAL(A);
    -- If variable A has value 2, signal B will get the real value 2.0.
```

```
type UNSIGNED is array (NATURAL range <>) of X01Z;
type X01Z_VECTOR is array (NATURAL range <>) of X01Z;
... UNSIGNED (TO_X01ZVECTOR (INT => 56, SIZE => 9)) ...
```

-- Since UNSIGNED and X01Z_VECTOR types have the same element
-- types (which is X01Z) and the same index types (which is NATURAL),
-- they are closely related types, and hence type conversions between
-- these two types are allowed.

```
type T1 is array (12 to 15) of INTEGER;
type T2 is array (3 downto 0) of INTEGER;
variable V1: T1;
variable V2: T2;
V1 := T1 (V2);
```

-- Type conversion is allowed between these two closely related types. Index
-- types can vary in index range and in their direction; however, number of
-- elements must be same.

2.82   *Type declaration*

A type represents a collection of values. An object that is declared to be of a certain type can assume values from those of the type. A type is declared using a type declaration.

Syntax

type_declaration

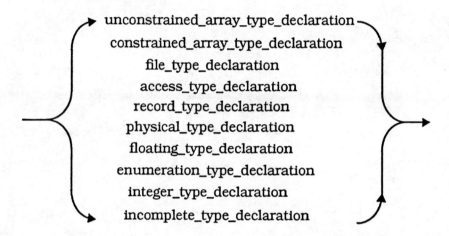

unconstrained_array_type_declaration
constrained_array_type_declaration
file_type_declaration
access_type_declaration
record_type_declaration
physical_type_declaration
floating_type_declaration
enumeration_type_declaration
integer_type_declaration
incomplete_type_declaration

Used In

"Architecture body" on page 24
"Block statement" on page 42
"Entity declaration" on page 95
"Generate statement" on page 120
"Package body" on page 159
"Package declaration" on page 162
"Process statement" on page 172
"Subprogram body" on page 202

Examples

type INSTR_SET **is** (NOOP, LOAD, MOVE, ADD, SUB, ANDP, NOTP);
    -- INSTR_SET is the name of an enumeration type that has a
    -- user-specified set of values.

signal DBUS: INSTR_SET;
    -- Signal DBUS can have values only from the type INSTR_SET and
    -- nothing else, that is, it can get one of the enumeration literals,
    -- LOAD, ADD, MOVE, etc., but can never get any other value, say '0'.

More examples of type declarations appear in their respective sections.

2.83 *Unconstrained array type declaration*

This declaration declares an array of an unspecified size, that is, only the array element type and the array index type are specified. The size of the array is specified when the unconstrained array type is used in an object (a variable, signal, or constant) declaration or in a subtype declaration.

The predefined unconstrained array types are STRING and BIT_VECTOR.

Syntax

unconstrained_array_type_declaration

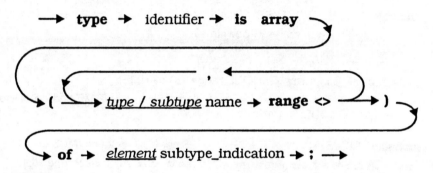

Used In

"Type declaration" on page 219

Examples

type REG_TYPE **is array** (NATURAL **range** <>) **of** STD_LOGIC;
>   -- REG_TYPE is an unconstrained array type. The element type of
>   -- REG_TYPE is STD_LOGIC and index type is NATURAL.

variable REG_A: REG_TYPE (0 **to** 31);
>   -- Size of array (32 elements) is specified in the variable declaration that uses
>   -- the unconstrained array type.

constant INIT_VAL: REG_TYPE := "001010";
>   -- Range of INIT_VAL is 0 to 5; it is determined from the value of the constant.

type FIFO_TYPE **is array** (POSITIVE **range** <>) **of** REG_TYPE(7 **downto** 0);
   -- FIFO_TYPE is another unconstrained array type, whose index is of type
   -- POSITIVE and each element is of type REG_TYPE(7 **downto** 0); in other
   -- words, FIFO_TYPE is a one-dimensional array of one dimension elements,
   -- with the first dimension of an unspecified range and the second dimension
   -- of range 7 downto 0, with each element of type STD_LOGIC.

signal MEM_B: FIFO_TYPE (1 **to** 1024);
   -- MEM_B is a one dimensional array of one-dimensional elements, of
   -- size 1024 * 8, each element of the array is of type STD_LOGIC.

subtype REG_TYPE_16 **is** REG_TYPE(31 **downto** 16);
   -- Note that the range can be specified as either an increasing range or a
   -- decreasing range.

variable REG_B: REG_TYPE_16;
   -- A variable with range 31 downto 16.

type OPCODE **is** (ADD, SUB, MUL, DIV);
type TIMING_T **is array** (OPCODE **range** <>) **of** TIME;
   -- TIMING_T ia an array type, each element is of type TIME and index is of
   -- type OPCODE.

constant TABLE: TIMING_T := (SUB => 25 ns, ADD => 23 ns, MUL => 56 ns);
   -- TABLE is a 3-element array with three indices, ADD, SUB, and MUL.
   -- The array index values must be a subrange of the index range of the type.

type MEMORY **is array** (NATURAL **range** <>, NATURAL **range** <>) **of** X01Z;
   -- MEMORY is a two-dimensional unconstrained array with each element of
   -- type X01Z.

subtype MY_MEM **is** MEMORY (0 **to** 7, 1023 **downto** 0);
   -- Size of array is specified in the subtype declaration for MY_MEM.

signal MEM_A: MY_MEM;
   -- MEM_A is a two-dimensional array signal.

2.84 *Use clause*

A use clause can be used to import items from a library or from a package into a design unit if specified just before a design unit. A use clause may also appear in any declarative region. In such a case, it imports items into the declarative region making them directly visible.

Syntax

use_clause

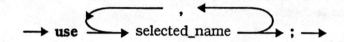

$$\rightarrow \textbf{use} \quad \overset{\longleftarrow}{\underset{\longrightarrow}{\quad}} \text{selected\_name} \overset{,}{\underset{\longrightarrow}{\quad}} ; \rightarrow$$

Used In

"Architecture body" on page 24
"Block statement" on page 42
"Configuration declaration" on page 71
"Design file" on page 90
"Entity declaration" on page 95
"Generate statement" on page 120
"Package body" on page 159
"Package declaration" on page 162
"Process statement" on page 172
"Subprogram body" on page 202

Examples

use DZX.LOGIC_ARITH.all;

    -- Imports all declarations from package LOGIC_ARITH that reside in the
    -- DZX design library.

use WORK.OR2;

    -- Imports the entity declaration for OR2 that resides in the WORK library.

use ATT.ATT_MVL.MVL;

    -- Imports only the type declaration for MVL from the package ATT_MVL that
    -- resides in the ATT design library.

```
use WORK.OR2, ATTLIB.XOR2, FPGA.AND2;

use STD.STANDARD.all;
    -- This use clause is implicitly present in all design units.

procedure WRITE_RESULTS is
    use STD.TEXTIO.all;
    variable RESULT_FILE: TEXT;
    variable L: LINE;
begin
    . . .
end procedure;
    -- The use clause makes all names declared in the TEXTIO package visible
    -- inside the procedure body.
```

2.85 *Variable assignment statement*

A variable assignment statement is a sequential statement that can appear within a process or a subprogram. When executed, it computes the value of the expression on the right-hand-side and assigns the value to a variable or to a set of variables (called an aggregate). The assignment occurs in zero simulation time, that is, the assignment is instantaneous.

Syntax

variable_assignment_statement

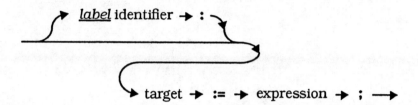

target

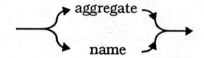

Used In

"Sequential statement" on page 188

Examples

variable BUFF: STRING (0 **to** 131);
BUFF := (**others** => '.');
   -- Each element of the BUFF array is assigned the '.' character.

V1: FAC := FAC + INDEX;
   -- The right-hand-side expression is evaluated, and the value is
   -- assigned to FAC. V1 is the label for the assignment statement.

(CIN, A3, A2, A1, A0) := STD_LOGIC_VECTOR'("10111");
    -- CIN, A3, A2, A1, and A0 have been assumed to be declared as variables of
    -- type STD_LOGIC. CIN gets the value '1', A3 gets '0', A2 gets '1', A1 gets '1'
    -- and A0 gets '1'. The left-hand-side of the assignment is an example of
    -- an aggregate.

UNSIG_A := TO_UNSIGNED (VALUE => RAD, SIZE => UNSIG_A'LENGTH);

J := 2;

CPU_BUS := (0 **to** 5 => '1', 6 | 8 | 11 => 'U', **others** => '0');
    -- An aggregate is assigned to the variable target.

2.86   *Variable declaration*

A variable declaration declares a variable and specifies the set of values (by the type information) that this variable can hold. The declaration may optionally specify an expression whose value becomes the initial value of the variable. If no expression is specified, then the initial value of the variable is the leftmost value of the variable type, that is, *variable_type*'LEFT; if the variable type is a composite type, then the initial value of each element of the composite type is the leftmost value of the corresponding element type.

A variable can be declared and used within a process or a subprogram. A variable declared outside of a process or a subprogram is called a shared variable; multiple processes can read and update such a shared variable simultaneously. Such a declaration must use the keyword **shared** explicitly.

A variable can be assigned a value using the variable assignment statement.

Syntax

variable_declaration

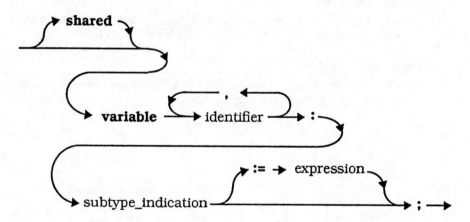

Used In

"Package declaration" on page 162
"Process statement" on page 172
"Subprogram body" on page 202

Examples

type STATE **is** (S0, S1, S2, S3, S4);
  -- An enumeration type.
variable CNT_STATE: STATE;
  -- Declares a variable CNT_STATE that can take only values belonging to the
  -- type STATE and the initial value of CNT_STATE is S0 (S0 is the leftmost
  -- value of the type STATE, that is, STATE'LEFT).

variable INDEX: NATURAL := 10;
  -- INDEX is a variable that can take values from the subtype NATURAL and
  -- its initial value is 10.

variable REG_BANK: BIT_VECTOR (63 **downto** 0) := (**others** => '1');
  -- REG_BANK is a one-dimensional array of bits, each element can take a
  -- value '0' or '1', and the initial value for all elements is '1'.

variable SUM, J : INTEGER **range** 0 **to** 50;
  -- SUM and J are variables of type INTEGER that are constrained to have
  -- integer values in the range 0 to 50. Their initial values are 0.

shared variable PV: BOOLEAN;
  -- PV is declared as a shared variable. This declaration can only appear
  -- outside of a process or a subprogram.

2.87   Wait statement

A **wait** statement is a sequential statement that can appear within a process statement or within a procedure body. The **wait** statement causes the process to suspend. If a **wait** statement is present in a procedure, then the process that called the procedure suspends.

A process can wait for an event to happen on any one of a set of signals, or it can wait for a time-out period, or it can wait until a certain condition becomes true, or it can wait for a combination of these.

Syntax

wait_statement

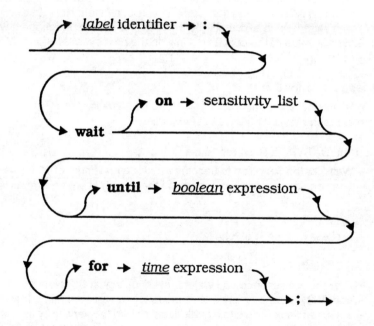

sensitivity_list

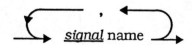

Used In

"Sequential statement" on page 188

Examples

wait on RESET, PRESET, CLOCK;

    -- On execution of this statement, the process suspends and waits for an
    -- event to occur on RESET, PRESET, or CLOCK signal.

wait for 20 ms;

    -- Causes the process to suspend for 20 ms.

wait until ACK = '1';

    -- Waits until the specified condition becomes true. Every time there is an
    -- event on ACK, the boolean expression is evaluated. Note that when this
    -- wait statement is executed, the process first suspends, irrespective of the
    -- current value of the condition. Only on the next event on signal ACK is the
    -- condition evaluated.

W1: **wait on** S1 **for** 5 ns;

    -- Waits until an event occurs on S1 or 5 ns elapses.
    -- W1 is the label for the wait statement.

wait until (DATA = FETCH) **for** 10 ns;

    -- Waits for the condition to become true for a maximum of 10 ns.
    -- The condition is checked every time there is an event on signal DATA.

wait;

    -- Process suspends indefinitely.

wait on CK **until** RST = '0';

    -- Process will suspend and wait for event on signal CK. Note that if both a
    -- sensitivity list and a condition is present in a wait statement, then the wait
    -- statement is sensitive to signals listed only in the sensitivity list.
    -- Process will resume execution only if condition is true, that is, RST is '0',
    -- when there is an event on CK.

                                                ❑

CHAPTER 3   *Predefined Environment*

This chapter gives examples of the predefined attributes and gives an overview of the predefined packages, STANDARD and TEXTIO. The set of reserved words in the language is also listed.

3.1   Predefined attributes

3.1.1   Type attributes

These are attributes that are associated with types.

T 'BASE

This attribute when applied to any type or subtype, returns its base type.

```
type MVL4 is ('X', '0', '1', 'Z');
subtype RESOLVED is WIRED_OR MVL4;
subtype NORMALIZED is REAL range -1.0 to +1.0;
```

```
type FIFO is array (NATURAL range <>) of STD_LOGIC_VECTOR(1 to 64);
subtype FIFO16 is FIFO(0 to 15);
```

```
RESOLVED'BASE          is MVL4
NORMALIZED'BASE        is REAL
FIFO16'BASE            is FIFO
```

Note that this attribute can only be used as a prefix of some other attribute, such as

```
NORMALIZED'BASE'ASCENDING          is TRUE
RESOLVED'BASE'LOW                  is 'X'
```

T 'LEFT, *T* 'RIGHT, *T* 'HIGH, *T* 'LOW

These attributes can be applied to any scalar type T, that is, integer, floating point, physical, or enumeration type. The attribute returns the appropriate bound of the type, as shown in these examples.

```
type WEEK is (MON, TUE, WED, THU, FRI, SAT, SUN);
type ABS_INT is range 1 to 27;
subtype HOLIDAY is WEEK range SUN downto SAT;
subtype DELAY_TIME is TIME range 0 ns to 100 ns;
```

```
WEEK'HIGH = WEEK'RIGHT            is SUN
WEEK'LOW = WEEK'LEFT             is MON
HOLIDAY'LEFT = HOLIDAY'HIGH      is SUN
HOLIDAY'RIGHT = HOLIDAY'LOW      is SAT
DELAY_TIME'LEFT                  is 0 ns
DELAY_TIME'HIGH                  is 100 ns
```

For ascending ranges,

```
T'HIGH = T'RIGHT
T'LOW = T'LEFT
```

For descending ranges,

```
T'HIGH = T'LEFT
T'LOW = T'RIGHT
```

T 'ASCENDING

This attribute when applied to any scalar type returns a value TRUE if the range is an ascending range, FALSE otherwise. From the previous examples,

HOLIDAY'ASCENDING   is FALSE
DELAY_TIME'ASCENDING   is TRUE

T 'IMAGE(*X*), *T* 'VALUE (*X*)

These attributes convert a value to a string representation and vice versa. 'IMAGE takes a scalar type value and generates a string representation for it, whereas 'VALUE takes a string representation of a scalar type value and generates its equivalent value representation. Here are some examples.

```
type TEST is {A, \A\, 'A'};              -- Enumeration type.
type NUMERIC is range 1 to 16;           -- Integer type.
type CAP is 0 to 5000
    units
        pf;
        nf = 1000 pf;
    end units;                           -- Physical type.
```

TEST'IMAGE (A)   is "a"
    -- For basic identifiers, the result is in lower case.
TEST'IMAGE('A')   is " 'A' "
    -- For character literals, the quotes are retained.
TEST'IMAGE(\A\)   is "\A\"
    -- For extended identifiers, backslashes are retained.
TEST'IMAGE(\A\\B\)   is "\A\\B\"
    -- If backslash character present in an extended identifier, then it is doubled.

NUMERIC'IMAGE(12)   is "12"
NUMERIC'IMAGE(1_4)   is "14"
    -- Underlines are deleted in a numeric type value.
NUMERIC'IMAGE(10)   can be "10" or "1e1"
    -- Language does not define whether it should be in decimal form or
    -- exponent form. If it appears in exponent form, then "e" is in lower case.

CAP'IMAGE(5 nf)   is "5000 pf"
    -- Value is always expressed in number of base units of the physical type.
CAP'IMAGE(20 pf)   is "20 pf"
    -- Units always appear in lower case.

TIME'IMAGE(5 ps)   is "5000 fs"
    -- For type TIME, value is always expressed in units of the smallest
    -- resolution time being used for simulation.

```
TIME'VALUE("5000 fs")          is 5000 fs
  -- Result may use any unit of type TIME.

CAP'VALUE("2000 pf")           can be 2 nf or 2000 pf

NUMERIC'VALUE("13")            is 13
TEST'VALUE("A")                is A
TEST'VALUE("\A\")              is \A\
```

The following is true of these attributes, except for a real type. If SV is a value of type SV_TYPE, then

```
SV = SV_TYPE'VALUE (SV_TYPE'IMAGE(SV))
```

T 'POS(*X*), *T* 'VAL(*N*), *T* 'SUCC(*X*), *T* 'PRED(*X*), *T* 'LEFTOF(*X*), *T* 'RIGHTOF(*X*)

All these attributes operate on a value of a discrete type (integer or enumeration type) or a physical type. X is an expression of type T. N is an integer value.

The 'POS attribute gives the position number of a value of the specified type, while 'VAL does the inverse, that is, gives the value corresponding to the position number of the specified type.

```
type FSM is (F1, F2, F3, F4, F5);
```

```
FSM'POS(F3)              is 2, which is its position number.
NATURAL'POS(20)         is 20
TIME'POS(2 ps)          is 2000. For a physical type, position
                        number is the number of base units.

FSM'VAL(4)              is F5
NATURAL'VAL(15)         is 15
FSM'VAL(2)              is F3
TIME'VAL(536)           is 536 fs
```

The 'SUCC attribute gives the value of the specified type whose position number is one greater than that of the value specified. The 'PRED attribute gives a value whose position number is one less than that of the value specified.

```
INTEGER'SUCC(15)        is 16
INTEGER'PRED(2)         is 1
FSM'SUCC(F4)            is F5
```

FSM'SUCC(F5)	is an error
FSM'PRED(F2)	is F1
FSM'PRED(F1)	is an error

The 'LEFTOF attribute returns a value of the specified type that is to the left of the specified value, while the 'RIGHTOF attribute returns a value that is to the right of the specified value.

```
subtype INDEX is INTEGER range 15 downto 0;
type CAP is 0 to 5000
    units
        pf;
        nf = 1000 pf;
    end units;
```

INDEX'LEFTOF(14)	is 15
INDEX'RIGHTOF(14)	is 13
INDEX'LEFTOF(15)	is an error
INDEX'RIGHTOF(0)	is an error
INDEX'SUCC(14)	is 15. Contrast with 'RIGHTOF.
INDEX'PRED(14)	is 13. Contrast with 'LEFTOF.
FSM'LEFTOF(F4)	is F3
FSM'RIGHTOF(F2)	is F3
CAP'SUCC(5 pf)	is 6 pf
CAP'PRED(5 pf)	is 4 pf
CAP'LEFTOF(5 nf)	is 4999 pf
CAP'RIGHTOF(10 pf)	is 11 pf
CAP'POS(7 pf)	is 7
CAP'VAL (2000)	is 2000 pf

3.1.2   Array attributes

The following attributes operate on constrained array objects or types. In all the following attributes, N specifies the dimension of the array and if not specified, defaults to 1.

A'LEFT [(N)], A'RIGHT [(N)], A'HIGH [(N)], A'LOW [(N)]

The 'LEFT attribute returns the left bound of the Nth index range of array A. The 'RIGHT attribute returns the right bound of the Nth index range of array A. The 'HIGH attribute returns the upper bound, while the 'LOW attribute returns the lower bound, of the Nth index range of the array.

```
type THREE_D is array (0 to 4, 1 to 15, 100 downto 75) of STD_LOGIC;
variable MEMO: THREE_D;
    -- MEMO is a three-dimensional array.
```

MEMO'LEFT	is 0. Left bound of 1st index range.
MEMO'RIGHT(2)	is 15
MEMO'HIGH(3)	is 100
MEMO'LOW(3)	is 75
MEMO'HIGH(1)	is 4
MEMO'LOW	is 0. Lower bound of 1st index range.

```
signal REG_FILE: THREE_D;
type SUB_THREE_D is array (10 downto 9, 100 downto 95, 50 to 60)
    of STD_LOGIC;
alias REGA: SUB_THREE_D is REG_FILE(1 to 2, 1 to 6, 100 downto 90);
```

REGA'LEFT	is 10. Left bound of 1st index range of alias, not of array.
REGA'LOW(2)	is 95
REGA'HIGH(3)	is 60

A'RANGE [(*N*)], *A*'REVERSE_RANGE [(*N*)]

The 'RANGE attribute returns the range of the Nth index range of array A. The 'REVERSE_RANGE attribute returns the reverse range, that is, if 'RANGE returns an ascending order, 'REVERSE_RANGE returns a descending order and vice versa.

```
variable AXE: BIT_VECTOR (0 to 63);
constant MAX: POSITIVE := 12;
type TWO_D_ARR is array (1 to MAX, 63 downto 0) of STD_LOGIC;
signal BOX: TWO_D_ARR;
```

AXE'RANGE	is "0 to 63". N is by default 1.
AXE'REVERSE_RANGE(1)	is "63 downto 0"
BOX'RANGE(2)	is "63 downto 0"
BOX'REVERSE_RANGE	is "12 downto 1"

A'LENGTH [(*N*)]

The 'LENGTH attribute returns the size of the Nth index range, that is, the number of values in this range.

    BOX'LENGTH   is 12. 1st index range by default.
    BOX'LENGTH(2)   is 64

A'ASCENDING [(*N*)]

The 'ASCENDING attribute returns the value TRUE if the Nth index range of the array A is an ascending range, otherwise it returns FALSE.

    BOX'ASCENDING   is TRUE. By default, 1st index range.
    BOX'ASCENDING(2)   is FALSE
    AXE'ASCENDING   is TRUE

3.1.3   Signal attributes

The following attributes can be applied to signals. For attributes that take an optional time value, the default value is 0 ns if not specified.

S'DELAYED [(*T*)], *S*'STABLE [(*T*)], *S*'QUIET [(*T*)], *S*'TRANSACTION [(*T*)]

The 'DELAYED attribute generates a new signal that is equivalent in waveform to signal S, but which is delayed by T time units.

The 'STABLE attribute generates a new BOOLEAN signal that has a value TRUE if an event has not occurred on signal S for T time units, else the value is FALSE.

The 'QUIET attribute also generates a BOOLEAN signal that has a value TRUE if signal has been quiet, that is, not active, for T time units.

It is important to differentiate a signal being active versus a signal having an event. A signal is said to be active whenever a value is assigned to the signal, irrespective of the old and new values of the signal. However, an event is said to occur on the signal if the signal is active and if the old value is different from the new value being assigned.

The 'TRANSACTION attribute generates a new signal of type BIT that toggles its value every time signal S becomes active, that is, every time a value is assigned to signal S.

Figure 3.1 shows an example of these four attributes.

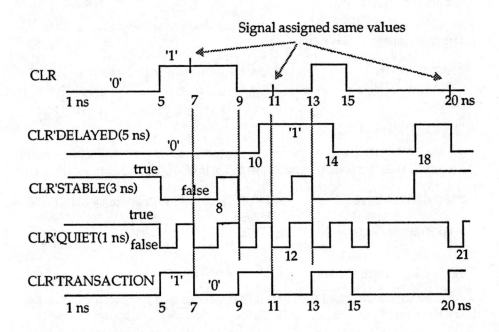

Figure 3.1: Signal attributes.

S'EVENT, *S*'ACTIVE, *S*'LAST_EVENT, *S*'LAST_ACTIVE, *S*'LAST_VALUE

The 'EVENT attribute returns a boolean value. It returns TRUE if an event just occurred on signal S, otherwise it is FALSE.

The 'ACTIVE attribute returns a boolean value of TRUE if signal S is active, otherwise it returns FALSE.

The 'LAST_EVENT attribute returns the time elapsed since the last event on signal S.

The 'LAST_ACTIVE attribute returns the time elapsed since the last time signal S was active.

The 'LAST_VALUE attribute returns the previous value of signal S, that is, the value before the last change on S.

```
CLR'EVENT at 5 ns          is TRUE
CLR'EVENT at 7 ns          is FALSE
CLR'EVENT at 15 ns         is TRUE
```

CLR'ACTIVE at 5 ns	is TRUE
CLR'ACTIVE at 7 ns	is TRUE
CLR'ACTIVE at 9 ns	is TRUE
CLR'ACTIVE at 10 ns	is FALSE
CLR'ACTIVE at 11 ns	is TRUE
CLR'LAST_EVENT at 8 ns	is 3 ns
CLR'LAST_EVENT at 13 ns	is 0 ns
CLR'LAST_EVENT at 20 ns	is 5 ns
CLR'LAST_ACTIVE at 8 ns	is 1 ns
CLR'LAST_ACTIVE at 9 ns	is 0 ns
CLR'LAST_ACTIVE at 10 ns	is 1 ns
CLR'LAST_ACTIVE at 20 ns	is 0 ns
CLR'LAST_VALUE at 7 ns	is '0'
CLR'LAST_VALUE at 9 ns	is '1'
CLR'LAST_VALUE at 12 ns	is '1'
CLR'LAST_VALUE at 14 ns	is '0'

S'DRIVING, S'DRIVING_VALUE

The 'DRIVING attribute gives a boolean value of FALSE if the current value of the driver for signal S in the current process is determined by a null transaction, otherwise it is TRUE (a *null transaction* is a transaction in which the value null is assigned to a signal). The 'DRIVING_VALUE attribute gives the current value of the driver for signal S in the current process.

These attributes can only be used within a process statement, a concurrent statement with an equivalent process statement, and within a subprogram that is in the declarative part of a process. It is an error to use this attribute with a port of mode **in**. Similarly, if the signal is a subprogram parameter, it must be of mode **out** or **inout**. The 'DRIVING_VALUE attribute must not be used when the 'DRIVING attribute has the value FALSE.

```
process
begin
    . . .
    if A = '0' then
        CAR <= null;          -- A null transaction is assigned to signal CAR.
    else
        CAR <= '1';
    end if;
```

. . .
end process;

If process executes say at time T, and if A has the value '0', then CAR'DRIVING is false at time T in this process. If A is not equal to '0', then CAR'DRIVING_VALUE at time T is '1' and CAR'DRIVING is true, assuming no other signal assignment to CAR exists in this process.

3.1.4   Named item attributes

These attributes are associated with any named items, such as, label, variable, signal, entity name, component name, etc., except local ports and generics of a component declaration.

E'SIMPLE_NAME

This attribute returns a string value for any name of a named item.

```
signal CLK: BIT;
type MC_STATE is (READY, WAITING, HOLD, RUNNING);
variable \Wait\: STD_LOGIC;
type ABC is ('A', 'B', 'C');
```

CLK'SIMPLE_NAME   is the string "clk"
   -- Simple names are always converted to lower case.

READY'SIMPLE_NAME   is the string "ready"

"+" 'SIMPLE_NAME   is the string "+"

"MOD" 'SIMPLE_NAME   is the string "mod"
   -- Operator symbols are converted to lower case.

\Wait\'SIMPLE_NAME   is the string "\Wait\"
'C' 'SIMPLE_NAME   is the string " 'C' "
   -- Case, single quotes, and backslashes are preserved for
   -- extended identifiers.

E'INSTANCE_NAME

This attribute gives a string that contains the hierarchical path for the named item starting from the root of the elaborated design hierarchy. It also includes the names of the instantiated design entities.

There are two kinds of instance names. A package-based path identifies items declared in a package, while a full-instance-based path identifies items in a design hierarchy. All characters in the instance name appear in lower case.

Examples are given in the following subsection.

E'PATH_NAME

This attribute returns a string that contains the hierarchical path name starting from the root of the elaborated design hierarchy, but this time it excludes the name of the instantiated design entities.

The syntax for the path name is very similar to that of the 'INSTANCE_NAME attribute except that the instantiated design unit does not appear. As in the previous attribute, all characters in the path name appear in lower case.

The following defines the syntax of a hierarchical instance name and path name.

instance_name

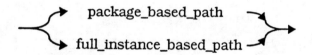

path_name

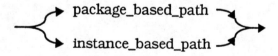

package_based_path

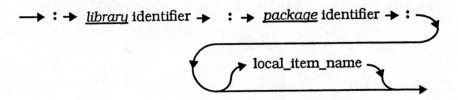

full_instance_based_path

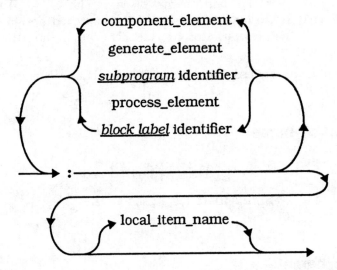

instance_based_path

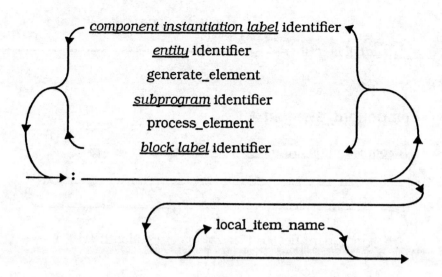

local_item_name

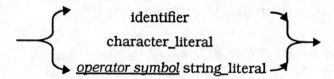

process_element

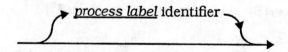

generate_element

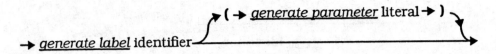

component_element

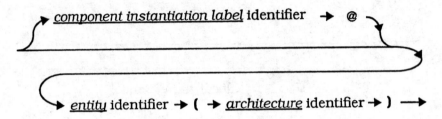

A component label is not needed in the component element if entity and architecture refer to a top-level design entity. If a process label does not exist, an empty string is used for process element.

Examples for 'INSTANCE_NAME attribute and 'PATH_NAME attribute are given below.

Following are examples of package-based paths:

```
BIT_ARITH'INSTANCE_NAME          is ":att:bit_arith:"
BIT_ARITH'PATH_NAME              is same as 'INSTANCE_NAME
    -- Package BIT_ARITH present in design library ATT.

LONG'INSTANCE_NAME          is ":dzx:bit_utils:long:"
LONG'PATH_NAME              is same as 'INSTANCE_NAME
    -- Subtype LONG declared in package BIT_UTILS present in library DZX.

TO_UX01'INSTANCE_NAME     is ":ieee:std_logic_1164:to_ux01:"
TO_UX01'PATH_NAME         is same as 'INSTANCE_NAME
    -- Function TO_UX01 declared in package STD_LOGIC_1164 present in
    -- library IEEE.
```

Following are examples of instance-based paths:

FULL_ADDER'INSTANCE_NAME   is ":full_adder(dataflow):"
FULL_ADDER'PATH_NAME   is ":full_adder:"

    -- Assuming that FULL_ADDER is the root design entity; otherwise
    -- a component label must also be specified for 'INSTANCE_NAME.

CARRY'INSTANCE_NAME   is ":full_adder (dataflow):carry"
CARRY'PATH_NAME   is ":full_adder:carry"

    -- Signal CARRY declared in the declarative part of
    -- the FULL_ADDER design entity.

NEXT_STATE'INSTANCE_NAME   is ":fa(str)::next_state"
NEXT_STATE'PATH_NAME   is ":fa::next_state"

    -- Variable NEXT_STATE declared in a process with no label in
    -- design entity FA(STR).

SC'INSTANCE_NAME   is ":fa(str):b1:b2:sc"
SC'PATH_NAME   is ":fa:b1:b2:sc"

    -- Signal SC declared in block B2 which is inside block B1 which is
    -- in architecture STR of entity FA.

C'INSTANCE_NAME   is ":fa(str):h1@ha(df):c"
C'PATH_NAME   is ":fa:h1:c"

    -- C is a signal declared in architecture DF of entity HA, which is
    -- instantiated in component label H1 in architecture STR of entity FA.

D_TMP'INSTANCE_NAME   is ":counter(str):ff(4):d_tmp"
D_TMP'PATH_NAME   is ":counter:ff(4):d_tmp"

    -- D_TMP is a signal declared in a generate statement labeled FF with a
    -- generate parameter value of 4 in architecture STR of entity COUNTER.

SUM'INSTANCE_NAME   is ":fft(seq):add_op:sum"
SUM'PATH_NAME   is ":fft:add_op:sum"

    -- Variable SUM declared in function ADD_OP that is present in
    -- architecture SEQ of entity FFT.

3.2   Package STANDARD

This package contains the predefined types and operations of the language.
The two context clauses

```
library STD, WORK;
use STD.STANDARD.all;
```

are implicitly present in each and every design unit and need not be specified explicitly. Therefore all items declared in package STANDARD are directly visible in every design unit. The items declared in this package are as follows.

BOOLEAN enumeration type

This type has the values, FALSE and TRUE.

BIT enumeration type

This type has the values '0' and '1'.

CHARACTER enumeration type

This type contains all the allowed characters in the language (8-bit coded character set from ISO 8859-1: 1987(E)) that include among others, upper-case letters, lower-case letters, digits, and other special characters.

SEVERITY_LEVEL enumeration type

This type contains four values, NOTE, WARNING, ERROR, and FAILURE. It is used in assertion and report statements.

INTEGER integer type

This type defines the range of integer values. Any implementation must support at least the range $-(2^{31}-1)$ to $+(2^{31}-1)$.

REAL floating-point type

This type defines the range of real numbers. Any implementation must support the range $-1.0E38$ to $+1.0E38$.

TIME physical type

This type defines the range of time and also defines the units of time. Any implementation must support the range, $-(2^{31} - 1)$ fs to $+(2^{31} - 1)$ fs. The other units of time are ps, ns, us, ms, sec, min, and hr.

DELAY_LENGTH physical subtype

Defines a subtype of type TIME that ranges from 0 to TIME'HIGH.

NOW impure function

The function returns the current simulation time.

NATURAL integer subtype

This is a subtype of type INTEGER with a range of 0 to INTEGER'HIGH.

POSITIVE integer subtype

This is a subtype of type INTEGER that ranges from 1 to INTEGER'HIGH.

STRING array type

This is an unconstrained array of characters, that is, of type CHARACTER.

BIT_VECTOR array type

This is an unconstrained array of bits, that is, of type BIT.

FILE_OPEN_KIND enumeration type

This type specifies the access mode for files, that is, READ_MODE, WRITE_MODE, or APPEND_MODE.

FILE_OPEN_STATUS enumeration type

This type specifies the return status of a file open operation. Allowed values for this type are

- OPEN_OK: open was successful
- STATUS_ERROR: file was already open
- NAME_ERROR: file not found
- MODE_ERROR: could not open file with specified access mode

'FOREIGN attribute

This attribute can only be associated with an architecture body or a subprogram. It is used to specify implementation-specific information for an architecture body or subprogram.

```
architecture A of E is
    attribute FOREIGN of A: architecture is "xrotate -d2 -f";
begin
    -- Architecture body is described by executing the implementation-
    -- specific information provided by the 'FOREIGN attribute.
    -- Consequently, the architecture body need not have any statements.
end architecture A;
```

Similarly for a subprogram,

```
procedure PRINT_LINE (A: STRING) is
    attribute FOREIGN of PRINT_LINE: procedure is "putline(A)";
begin
    -- No body needed since call to PRINT_LINE causes the
    -- implementation-specific info to be executed.
end procedure PRINT_LINE;

package P is
    function ATOI (S: STRING) return INTEGER;
    attribute FOREIGN of ATOI: function is "/bin/sh atoi";
end package P;
```

3.3   Package TEXTIO

This package contains subprograms related to reading and writing from human readable ASCII files. To use information from this package, a design unit must contain a use context clause explicitly, such as

use STD.TEXTIO.**all;**

The complete listing for the package declaration follows.[1]

---

1. Reprinted from IEEE Std 1076-1993, IEEE Standard VHDL Language Reference Manual, copyright © 1993 by the Institute of Electrical and Electronics Engineers, Inc.

```
package TEXTIO is
     -- Type definitions for text I/O:
     type LINE is access STRING; -- A line is a pointer to a
                                 -- STRING value.
     type TEXT is file of STRING; -- A file of variable-length
                                  -- ASCII records.
     type SIDE is (RIGHT, LEFT); -- For justifying output data
                                 -- within fields.
     subtype WIDTH is NATURAL; -- For specifying widths of
                               -- output fields.

     -- Standard text files:
     file INPUT: TEXT open READ_OPEN is "STD_INPUT";
     file OUTPUT: TEXT open WRITE_OPEN is "STD_OUTPUT";

     -- Input routines for standard types:
     procedure READLINE (file F: TEXT; L: out LINE);

     procedure READ (L: inout LINE; VALUE: out BIT;
                     GOOD: out BOOLEAN);
     procedure READ (L: inout LINE; VALUE: out BIT);

     procedure READ (L: inout LINE; VALUE: out BIT_VECTOR;
                     GOOD: out BOOLEAN);
     procedure READ (L: inout LINE; VALUE: out BIT_VECTOR);

     procedure READ (L: inout LINE; VALUE: out BOOLEAN;
                     GOOD: out BOOLEAN);
     procedure READ (L: inout LINE; VALUE: out BOOLEAN);

     procedure READ (L: inout LINE; VALUE: out CHARACTER;
                     GOOD: out BOOLEAN);
     procedure READ (L: inout LINE; VALUE: out CHARACTER);

     procedure READ (L: inout LINE; VALUE: out INTEGER;
                     GOOD: out BOOLEAN);
     procedure READ (L: inout LINE; VALUE: out INTEGER);

     procedure READ (L: inout LINE; VALUE: out REAL;
                     GOOD: out BOOLEAN);
     procedure READ (L: inout LINE; VALUE: out REAL);

     procedure READ (L: inout LINE; VALUE: out STRING;
                     GOOD: out BOOLEAN);
     procedure READ (L: inout LINE; VALUE: out STRING);
```

```
                    procedure READ (L: inout LINE; VALUE: out TIME;
                                    GOOD: out BOOLEAN);
                    procedure READ (L: inout LINE; VALUE: out TIME);

                    -- Output routines for standard types:
                    procedure WRITELINE (file F: TEXT; L: in LINE);

                    procedure WRITE (L: inout LINE; VALUE: in BIT;
                        JUSTIFIED: in SIDE := RIGHT; FIELD: in WIDTH := 0);
                    procedure WRITE (L: inout LINE; VALUE: in BIT_VECTOR;
                        JUSTIFIED: in SIDE := RIGHT; FIELD: in WIDTH := 0);
                    procedure WRITE (L: inout LINE; VALUE: in BOOLEAN;
                        JUSTIFIED: in SIDE := RIGHT; FIELD: in WIDTH := 0);
                    procedure WRITE (L: inout LINE; VALUE: in CHARACTER;
                        JUSTIFIED: in SIDE := RIGHT; FIELD: in WIDTH := 0);
                    procedure WRITE (L: inout LINE; VALUE: in INTEGER;
                        JUSTIFIED: in SIDE := RIGHT; FIELD: in WIDTH := 0);
                    procedure WRITE (L: inout LINE; VALUE: in REAL;
                        JUSTIFIED: in SIDE := RIGHT; FIELD: in WIDTH := 0;
                        DIGITS: in NATURAL := 0);
                    procedure WRITE (L: inout LINE; VALUE: in STRING;
                        JUSTIFIED: in SIDE := RIGHT; FIELD: in WIDTH := 0);
                    procedure WRITE (L: inout LINE; VALUE: in TIME;
                        JUSTIFIED: in SIDE := RIGHT; FIELD: in WIDTH := 0;
                        UNIT: in TIME := ns);

                    -- File position predicate:
                    -- function ENDFILE (file F: TEXT) return BOOLEAN;
                end TEXTIO;
```

Here is an example of a process that shows the usage of READ and READLINE procedures.

```
process
    file VEC_FILE: TEXT open READ_MODE is "gtc.in_vec";
        -- File "gtc.in_vec" contains a TIME value, a BIT value, a 3-bit
        -- BIT_VECTOR value, and an INTEGER value.
    variable TIME_TO_APPLY: TIME;
    variable RESET: BIT;
    variable IFRM: BIT_VECTOR(0 to 3);
    variable SHIFT_BITS: INTEGER;
    variable BUF: TEXT;
begin
    while not ENDFILE (VEC_FILE) loop
        READLINE (VEC_FILE, BUF);        -- Read one line from file.
        READ (BUF, TIME_TO_APPLY);       -- Read first field in line.
```

```
            READ (BUF, RESET);                  -- Read second field.
            READ (BUF, IFRM);                   -- Read third field.
            READ (BUF, SHIFT_BITS);             -- Read fourth field.
            -- Do whatever computations here with the above read values.
            . . .
        end loop;
    end process;
```

Here is an example of a process that shows the usage of WRITE and WRITELINE procedures.

```
        signal SERIAL_IN: BIT;
        signal PARALLEL_OUT: BIT_VECTOR (0 to 7);
        signal RESET, PARITY_ERROR: BOOLEAN;
    . . .
    process
        file OUT_FILE: TEXT open WRITE_MODE is "ser_par.out";
        variable BUF: LINE;
    begin
        wait until CLOCK = '1';             -- Wait for rising clock edge.

        WRITE (BUF, STRING'("At "));        -- Write into buffer.
        WRITE (BUF, NOW);                   -- Write into buffer.
        WRITE (BUF, STRING'(": SERIAL_IN = "));
        WRITE (BUF, SERIAL_IN);
        WRITE (BUF, STRING'(", RESET = "));
        WRITE (BUF, RESET);
        WRITE (BUF, STRING'(", PARALLEL_OUT = "));
        WRITE (BUF, PARALLEL_OUT);
        WRITELINE (OUT_FILE, BUF);          -- Write buffer to file.
    end process;
```

The file "ser_par.out" will have lines such as

```
At 20 ns, SERIAL_IN = 1, RESET = FALSE, PARALLEL_OUT = 01000100
At 40 ns, SERIAL_IN = 0, RESET = FALSE, PARALLEL_OUT = 01011100
```

3.4   Reserved words

This section lists the reserved words of the language.[2] These cannot be used as basic identifiers.

abs	access	after	alias
all	and	architecture	array
assert	attribute		
begin	block	body	buffer
bus			
case	component	configuration	constant
disconnect	downto		
else	elsif	end	entity
exit			
file	for	function	
generate	generic	group	guarded
if	impure	in	inertial
inout	is		
label	library	linkage	literal
loop			
map	mod		
nand	new	next	nor
not	null		
of	on	open	or
others	out		
package	port	postponed	procedure
process	pure		

---

2. Reprinted from IEEE Std 1076-1993, IEEE Standard VHDL Language Reference Manual, copyright © 1993 by the Institute of Electrical and Electronics Engineers, Inc.

range	**record**	**register**	**reject**
rem	**report**	**return**	**rol**
ror			
select	**severity**	**shared**	**signal**
sla	**sll**	**sra**	**srl**
subtype			
then	**to**	**transport**	**type**
unaffected	**units**	**until**	**use**
variable			
wait	**when**	**while**	**with**
xnor	**xor**		

❏

CHAPTER 4   *Changes from VHDL '87*

This chapter presents the changes that were introduced to the 1076-1987 version of the language. Also included is a section on portability of models from 1076-1987 to 1076-1993 versions of the language.

4.1   *VHDL'93 features*

This section lists the salient changes that have been incorporated into the 1076-1993 version of the language. *In order to write models that are 1076-1987 compatible, these features must be avoided.* In addition to those listed below, 100+ ambiguities in the 1076-1987 version of the language were resolved.

1.   File is the fourth object class (in addition to variable, constant, and signal).

2.   A constant expression is allowed as an actual in a port map.

3.   Shared variables are allowed.

4.   The notion of a group has been introduced. A group is a collection of
     named items. A group template declaration is used to declare a group
     template, that is, the class of named items that form a group. A group
     declaration is used to declare a group.

5.   A new attribute 'FOREIGN is declared. This attribute can be used in an
     architecture body or a subprogram to link in non-VHDL models.

6.   The syntax has been made more uniform across a number of constructs.

```
component C is
  . . .
end component C;

process ( . . . ) is
  . . .
end process;

A: block ( . . . ) is
  . . .
end block A;

architecture A of E is
  . . .
end architecture A;

procedure P is
  . . .
end procedure P;

function F ( . . . ) is
  . . .
end function F;

entity E is
  . . .
end entity E;

configuration C of E is
  . . .
end configuration C;

package PK is
  . . .
end package PK;
```

```
package body PK is
    . . .
end package body PK;
```

7.   Any sequential statement can be labeled. For example,

```
L1: if A = B then        -- L1 is the if statement label.
      C := D;
end if L1;                -- Label can optionally appear at the end.

L2: SUM := (A xor B) xor C;
```

8.   Functions can be designated as pure or impure. A pure function is one that returns the same value for multiple calls to the function with the same set of parameter values; an impure function may potentially return a different value when called multiple times, even with the same parameter values.

9.   The notion of signature has been introduced. This can be used to explicitly identify overloaded subprograms and overloaded enumeration literals. The signature explicitly specifies the parameter and result profile.

10.   File operations that are implicitly defined when a file type is declared have been redefined. These are FILE_OPEN, FILE_CLOSE, READ, WRITE, and ENDFILE.

11.   Syntax of file declaration has been redefined.

12.   An alias can be specified for any named item, that is, an alias is not restricted to just objects. However, labels, and loop and generate parameters cannot be aliased.

13.   An attribute can be specified for a literal, units, group, and file.

14.   The xnor operator and shift and rotate operators have been defined.

15.   The results of a concatenation operator have been elaborated.

16.   The report statement has been introduced. It is very similar to an assertion statement but without the assert expression.

17.   A pulse rejection window can be specified when inertial delay is used in a signal assignment.

18.   A value of unaffected can be assigned to a signal to indicate no change to the value of the driver.

19. A process can be marked as a postponed process. A postponed process executes only at the end of a time step, that is, after all the deltas of a time step.

20. Similarly, a concurrent assertion statement, a concurrent procedure call, and a concurrent signal assignment statement can be marked as postponed.

21. Direct instantiation is allowed, that is, in a component instantiation statement, an entity-architecture pair or a configuration can be directly instantiated.

22. Incremental binding is allowed. For example, a configuration specification in an architecture body may specify the binding to a design entity without specifying the port map and the generic map, while a configuration declaration may be used later to specify the port map and the generic map.

23. The generate statement can have a declarative part.

24. The character set has been extended to include a number of other special characters.

25. Extended identifiers have been defined. An extended identifier is a sequence of characters written between two backslashes.

26. A bit string literal represents a sequence of bits. The type of literal need not necessarily be BIT_VECTOR; the type is determined from the context in which the literal appears.

27. The following predefined attributes have been added:

'ASCENDING, 'IMAGE, 'VALUE, 'DRIVING, 'DRIVING_VALUE, 'SIMPLE_NAME, 'INSTANCE_NAME, 'PATH_NAME

The following predefined attributes have been deleted:

'STRUCTURE, 'BEHAVIOR.

28. In package STANDARD, the following have been added:

- DELAY_LENGTH physical subtype,
- FILE_OPEN_KIND enumeration type,
- FILE_OPEN_STATUS enumeration type, and
- 'FOREIGN attribute declaration.

29.   The semantics of subprograms in the TEXTIO package have been elaborated.

4.2   *Portability from VHDL'87*

This section describes some of the features changed in 1076-1993 that can cause models written in 1076-1987 to be non-portable.

1.   Enumeration type CHARACTER has a much larger set. Therefore, code that depended on 'HIGH, 'RIGHT of type CHARACTER may encounter some changes.

2.   The operation performed by a concatenation operator has been elaborated.

3.   File type declaration has been redefined, and its associated implicit file operations have also been redefined.

4.   'STRUCTURE and 'BEHAVIOR attributes are no longer present in the language. Therefore, any code that uses these attributes will need to be changed.

5.   New reserved words have been added to the language. Therefore, if these were used as identifiers, the identifiers will have to be changed.

❑

Bibliography

Following is a list of suggested readings and books on the language. The list is not intended to be comprehensive.

1. Armstrong, J. R., *Chip-level Modeling with VHDL*, Englewood Cliffs, NJ: Prentice Hall, 1988.

2. Armstrong, J.R. et al., *The VHDL Validation Suite*, Proc. 27th Design Automation Conference, June 1990, pp. 2-7.

3. Ashenden, P.J., *The Designers Guide to VHDL*, Morgan Kaufmann, 1994.

4. Ashenden, P.J., *The VHDL Cookbook*, The University of Adelaide, Australia, 1990.

5. Baker, L., *VHDL Programming with Advanced Topics*, John Wiley and Sons, Inc., 1993.

6. Barton, D., *A First Course in VHDL*, VLSI Systems Design, January 1988.

7. Berge, J-M., et al., *VHDL Designer's Reference*, Kluwer Academic, 1992.

8. Berge, J-M., et al., *VHDL'92*, Kluwer Academic, 1993.

9. Bhasker, J., *A VHDL Primer*, Englewood Cliffs, NJ: Prentice Hall, 1992.

10. Bhasker, J., *Process-Graph Analyzer: A Front-end Tool for VHDL Behavioral Synthesis*, Software Practice and Experience, vol. 18, no. 5, May 1988.

11.   Bhasker, J., *An Algorithm for Microcode Compaction of VHDL Behavioral Descriptions*, Proc. 20th Microprogramming Workshop, December 1987.

12.   Coelho, D., *The VHDL Handbook*, Boston: Kluwer Academic, 1988.

13.   Coelho, D., *VHDL: A Call for Standards*, Proc. 25th Design Automation Conference, June 1988.

14.   Farrow, R., and A. Stanculescu, *A VHDL Compiler based on Attribute Grammar Methodology*, SIGPLAN 1989.

15.   Gilman, A.S., *Logic Modeling in WAVES*, IEEE Design & Test of Computers, June 1990, pp. 49-55.

16.   Hands, J.P., *What is VHDL?* Computer-Aided Design, vol. 22, no. 4, May 1990.

17.   Harr, R., and A. Stanculescu (eds.), *Applications of VHDL to Circuit Design*, Boston: Kluwer Academic, 1991.

18.   Hines, J., *Where VHDL fits within the CAD Environment*, Proc. 24th Design Automation Conference, 1987.

19.   *IEEE Standard VHDL Language Reference Manual, Std 1076-1987*, IEEE, NY, 1988.

20.   *IEEE Standard VHDL Language Reference Manual, Std 1076-1993*, IEEE, NY, 1993.

21.   *IEEE Standard 1076 VHDL Tutorial*, CLSI, Maryland, March 1989.

22.   *IEEE Standard Interpretations: IEEE Std 1076-1987, IEEE Standard VHDL Language Reference Manual*, IEEE, 1992.

23.   *IEEE Standard Multivalue Logic System for VHDL Model Interoperability (Std_Logic_1164)*, Std 1164-1993, IEEE, 1993.

24.   Kim, K., and J. Trout, *Automatic Insertion of BIST Hardware using VHDL*, Proc. 25th Design Automation Conference, 1988.

25.   Leung, *ASIC System Design with VHDL*, Boston: Kluwer Academic, 1989.

26.   Lipsett, R., et. al., *VHDL: Hardware Description and Design*, Boston: Kluwer Academic, 1989.

27. Moughzail, M., et. al., *Experience with the VHDL Environment*, Proc. 25th Design Automation Conference, 1988.

28. Perry, D., *VHDL*, New York: McGraw Hill, 1991.

29. *Military Standard 454*, 1988 US Government Printing Office.

30. Navabi, Z., *VHDL Analysis and Modeling of Digital Systems*, McGraw Hill, 1993.

31. Saunders, L., *The IBM VHDL Design System*, Proc. 24th Design Automation Conference, 1987.

32. Schoen, J.M., *Performance and Fault Modeling with VHDL*, Englewood Cliffs, NJ: Prentice Hall, 1992.

33. Ward, P.C., and J. Armstrong, *Behavioral Fault Simulation in VHDL*, Proc. 27th Design Automation Conference, June 1990, pp. 587-593.

❑

Index

❏